Elmer G. Willyoung

Electrical and Scientific Instruments and Apparatus

November 20th, 1899

ELECTRICAL

AND

SCIENTIFIC INSTRUMENTS

AND APPARATUS.

NOVEMBER 20th, 1899.

ELMER G. WILLYOUNG,
82-84 Fulton Street,
New York.

NOTICE.

In ordering from this catalogue give both catalogue number and designation of item.

Remit by New York Draft, Money Order or Express. Add 25 cents to face of all checks, except on New York or Philadelphia, to cover the collection charge made by the N. Y. Clearing House.

Customers unknown to us must give satisfactory New York or Philadelphia reference or send cash with order. Or, we will ship C. O. D. provided cash remittance be made of at least ⅓ the amount of order ; but *no orders less than $5.00 will be shipped C. O. D.*

Charges for collection and return of money on C. O. D. shipments must be borne by the customer.

Goods ordered sent by mail must be prepaid and the postage included in remittance.

Packing will be charged at cost. All goods will be packed with the utmost care, but we assume no responsibility for breakage or other damage after the goods leave our hands. We guarantee all goods strictly as represented.

PREFATORY.

TO my friends, customers and the scientific and technical public generally I desire to state that I have succeeded to the business established and built up by Messrs. WILLYOUNG & Co., Philadelphia, retaining all the "good will" of the same, and have removed the headquarters of this business to New York City at the address below. I have also made a close alliance with Messrs. FOOTE, PIERSON & Co , successors to the manufacturing department of the late firm of E. S. GREELEY Co., by which they pass over to me all the "good will," etc., of such measuring instrument business as they have been doing. By the terms of this arrangement all "Willyoung" apparatus as also all the "Greeley" measuring instruments will be made in the shops of FOOTE, PIERSON & Co., under my immediate direction and supervision and according to my design. To take care of this business the already extensive shops of this firm have been greatly added to in floor space during the past three months, while a large number of the most recent high class tools have also been installed. The shop and laboratory have been put in charge of those holding the same positions under WILLYOUNG & Co.

All exclusive sales agencies for "Willyoung" apparatus have been abolished, and such apparatus may now be obtained direct from me or from the trade generally. I believe this will be of real advantage to the public. There are often questions arising, either before or after a sale is made, which a salesman cannot answer—technical questions which he cannot be expected to answer. He must refer them to the inventor, designer or manufacturer. This means delay, possible errors due to faulty transmission of the information through a second party, and expense, on account of this extra and superfluous correspondence, which, in the long run, the customer must certainly pay for. Should the purchaser write direct to the manufacturer, but purchase of the sales agent, there is certain to result confusion and annoyance all around.

Having been largely active in the beginning and development of fine instrument manufacturing in America, and having been associated with it for a number of years past, I feel that myself and my apparatus is now so well known as to require here no special sounding of trumpets.

All the instruments and apparatus in this catalogue have been very thoroughly revised during the Summer past, and every possible improvement as regards construction and convenience in manipulation has been effected. The entire line of instruments, therefore, is thoroughly up to date.

This catalogue has been prepared with great care and it is believed that the errors, if any, are extremely few in number. So far as possible the various instruments are brought under a logical classification. Each class is prefaced by succinct information as to the place of this class in the general scheme of instrumental work, and the general method of use required by the features which differentiate the individual types under a general group. It is believed that this information will be valued and will be found extremely useful in facilitating intelligent selection on the part of many customers.

The manufacture of two grades of apparatus, viz., Grade A and Grade B, has met with such unqualified approval on the part of the public that it is continued.

Grade A applies to apparatus finished throughout in the best possible manner. Apparatus of Grade B is guaranteed equal to Grade A in all essential respects (material, general workmanship and accuracy of adjustment), but does not present the same highly polished and beautiful exterior. As most instruments soon show the effect of use, Grade B is very popular and is purchased in preference to Grade A in the large majority of cases.

My facilities for the manufacture of all kinds of special apparatus, whether from customers or my own design, cannot be surpassed.

A force of men is continually engaged in making repairs. I can do such work reasonably and promptly. Many instruments now useless could be given an effective lease of life with a little alteration.

My system of data relating to instruments and measurements has been carefully and accurately kept for many years back. There are few instruments or measurements ever suggested which I cannot locate. This information is at the service of my customers.

I make my instruments as well as I know how; study them carefully always. But I do not know it all and I sometimes make mistakes. I welcome suggestions or criticisms and know that such will enable me to turn out a better product.

A picture always shows more than a description, and I would like to have every instrument illustrated; but this is commercially impossible. I do have, however, photographs of nearly everything made by me, and I shall be glad to send one to any interested party to whom the catalogue description is inadequate.

ELMER G. WILLYOUNG.

New York, November 20th, 1899.

——— ——

KEY TO REFERENCES USED.

C. & P.	Carhart & Patterson's "Electrical Measurements."
Ayrton.	Ayrton's "Practical Electricity," 3d Ed.
Palaz.	"Treatise on Industrial Photometry," by Palaz ; translated from the French by Patterson.
S. & G.	Stewart & Gee's "Elementary Practical Physics," Vol. II.
Henderson.	Henderson's "Practical Electricity and Magnetism."
Gray.	"Absolute Measurements in Electricity and Magnetism," two volumes, by A. Gray.
Phil. Mag.	"London and Edinburgh Philosophical Magazine and Journal of Science."
Kempe.	Kempe's "Hand-book of Electrical Testing." Fifth Edition

(4) The complete investigation of a battery so as to bring out its rate of polarization and recovery therefrom.

(5) The determination of capacities of insulated cables — this is a most important test in Telephone Cables, particularly when the speed and clearness of such is greatly weakened by too great capacity.

Uses of Paper Condensers — Condensers of this kind, when properly made, will stand easily at least 500 volts and occupy a space of about 41 cu. ins. per microfarad.

Some of their general applications are:

(1) To afford opportunities for students' exercises in measuring capacities by various methods.

(2) For studying the effects of residual charge and loss of energy in the dielectric of a condenser.

(3) For comparing electromotive forces, by charging the condenser and discharging it through a ballistic galvanometer.

(4) Where an electrometer is used to measure potentials in a circuit that is closed intermittently, leakage in the electrometer would introduce error into the result. A paper condenser, joined in parallel with the electrometer will almost entirely obviate the difficulty.

(5) To use in connection with induction coils, where the latter have not a variable condenser.

(6) For illustrating the theory of alternating currents ; this cannot be taught successfully without condensers to illustrate many of the important theorems. The general relations between current, E. M. F., inductance and capacitance in a circuit of resistance, capacity and self-induction, is exceedingly important to get fixed in students' minds, but very difficult without condensers of suitable capacity for actual work in the laboratory. One needs at least ten microfarads, of Adjustable Condenser for such purpose.

Another obvious application of paper condensers, but one seldom employed, though it should find favor, is their use.

A—To change a Direct Current of Low Potential into an Oscillatory Current of any Frequency of Potential.

B—To raise a Low Potential to a High Potential by means of Resonance.

C—To produce a Rotary Field from a Single Phase Circuit.

Condensers for Telephone and Telegraph Purposes— We make all forms and capacities of condenser both mounted and unmounted for use in telephone and telegraph work where they are used.

(1) For balancing lines in quadruplex and multiplex work.

2) In telephonic transmission for enabling A. C. coils to be used on open circuited lines.

(3) In general for " killing " the spark around relay and sounder " breaks," thus maintaining reliable and clean contacts.

MANUFACTURE OF THE WILLYOUNG CONDENSERS.

The mode of manufacture of the Willyoung Condensers is unique and essentially as follows. The materials, mica or paper and tin-foil, first being cut to size, are stacked up in alternate layers of dielectric and tin-foil. This stack is gently clamped together and heated in an oven for a long time to drive out moisture. The entire stack is then transferred to a hot bath of a special insulating compound and cooked for another considerable time, after which the whole affair is placed in a vacuum-chamber. This insures the thorough saturation of the materials by the insulating compound and the complete exclusion of air from the finished condenser.

The practical results of this method of construction are :

(1) NO BREAK-DOWN. Condenser break-downs are usually due to static electrification of the enclosed air which, bombarding to and fro with the alternating circuit pulsations, softens the insulation, wastes power and finally ruins the condenser. As a rule, any but fine mica condensers become so hot in five or ten minutes, on anything over 200 or 300 volts, alternating, as to liquefy the insulation and allow the condenser-sheets to float away. By excluding air, as above mentioned, the Willyoung Condensers are free from all such faults.

(2) NEGLIGIBLE HEATING.

(3) SMALL ABSORPTION.

(4) GREAT DIELECTRIC STRENGTH. The Willyoung Condensers, either mica or paper, are guaranteed to stand 500 and 1,000 volt circuits continuously. Actually, voltages several times these values, may be put upon the condensers for considerable times without risk.

WILLYOUNG PAPER CONDENSERS.

SINGLE VALUE, ACCURATE TO 1%.

W4000. 1-3 Micro-farad Capacity			*$8 00	+$10 00
W4001. 1-2	"	"	9 00	11 25
W4002. 1	"	"	10 00	12 50
W4003. 5	"	"	20 00	25 00
W4004. 10	"	"	30 00	37 50

W4000 to 4004 are mounted in polished cherry cases.

‡ WILLYOUNG ADJUSTABLE CONDENSERS.
(PAPER.)

The Adjustable Condenser, invented by Mr. Willyoung, is a distinct advance in condenser construction, and fills a long felt want. It enables, for the first time in the history of the art, capacities to be varied at will

* Guaranteed to stand 500 volts alternating current.
+ Guaranteed to stand 1000 volts alternating current.
‡ The Willyoung Adjustable Condensers have paper as a dielectric. Adjustable Mica Condensers of this type and design will be quoted when desired.

and instantly, just as has been done for a long time past with resistances. It is as much superior to the "plug in" and "plug out" forms of condenser as the modern "lever-switch" rheostat is superior to the old "plug" or "binding post" types used when electric lighting systems were first established. Many experiments impossible with "plug" condensers, on account of the impossibility of carrying in the memory the effects produced by *slight* variations of capacity, become easy with the Adjustable Condenser.

The plan of connections used in the Adjustable Condenser is similar to the arrangement of an English Dial Resistance Set in which each general division or unit has ten subdivisions. The members of any group may be thrown in or out of multiple among themselves by simple movement of the proper switch.

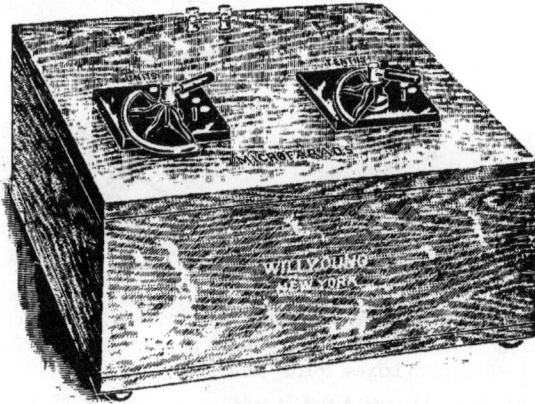

Fig. 2. W4011.

IMPROVED SWITCH.

Our first form of switch having occasionally given trouble by reason of defective contacts we have recently devised a new and improved form in which all defects have been eliminated and absolutely perfect contact at all times assured.

W4008. **1 Micro-farad Capacity**.................*$35 00   †$45 00
  10 sections of $\frac{1}{10}$ M. F. each.
W4009. **10 Micro-farads Capacity**..... .... 42 50   52 50
  10 sections of 1 M. F. each.
W4010. **100 Micro-farads Capacity**.... .... 300 00   375 00
  10 sections of 10 M. F. each.
W4011. **10 Micro-farads Capacity** 60 00   75 00
  10 sections of $\frac{1}{10}$ M. F. each and 9 sections of 1 M.F. each.
W4012. **100 Micro-farads Capacity**............. 350 00   425 00
  10 sections of $\frac{1}{10}$ M. F. each, 9 sections of 1 M. F. each
  and 9 sections of 10 M. F. each.
W4013. **100 Micro-farads Capacity**............. 325 00   400 00
  10 sections of 1 M F. each and 9 sections of 10 M.F.each.

---

* Guaranteed to stand 500 volts alternating current.
† Guaranteed to stand 1000 volts alternating current.

W4014. **X-Ray Adjustable Condenser**.......... ...... $30 00

Constructed especially for use with Induction Coils. Mounted in neatly finished cherry case. Should not be used on *alternating* circuits of more than *100 volts*. About 5 Micro-farads total.

Condensers W4008 to W4013 inclusive, are mounted in substantial wood cases and all values are accurate within 3%.

CONDENSERS FOR TELEGRAPH, TELEPHONE, ELECTRIC LIGHTING AND GENERAL EXPERIMENTAL PURPOSES.

Fig. 3. W4020 to W4034.

Guaranteed to stand 500 volts alternating current continuously without injury, change or injurious heating, and to be correct within 3%. Mounted in shellaced white pine boxes, neatly finished.

W4020.	**Single Valued Condenser**; capacity				$\frac{1}{50}$ M. F	......$ 3 00
W4021.	"	"	"	"	$\frac{1}{10}$	"	...... 4 50
W4022.	"	"	"	"	$\frac{2}{10}$	"	...... 5 00
W4023.	"	"	"	"	$\frac{1}{3}$	"	...... 5 50
W4024.	"	"	"	"	$\frac{1}{2}$	"	...... 6 00
W4025.	"	"	"	"	1	"	...... 10 00
W4026.	"	"	"	"	2	"	...... 12 50
W4027.	"	"	"	"	3	"	...... 15 00
W4028.	"	"	"	"	4	"	...... 17 50
W4029.	"	"	"	"	5	"	...... 20 00
W4030	"	"	"	"	6	"	...... 22 00
W4032.	"	"	"	"	8	"	...... 26 00
W4034.	"	"	"	"	10	"	...... 30 00

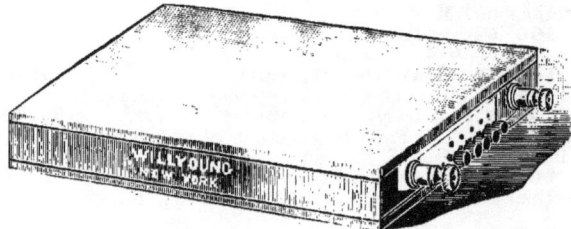

Fig. 4. W4040 to W4049.

W4040.	**Subdivided Multiple Condensers,** 5 sec's; tot. cap.					1 M.F.	$12 00
W4041.	``	``	``	``	``	``	2	``	14 50
W1042.	``	``	``	``	``	``	3	``	17 00
W4043.	``	``	``	``	``	``	4	``	19 50
W4044.	``	``	``	``	``	``	5	``	22 00
W4045.	``	``	``	``	``	``	6	``	24 00
W4047.	``	``	``	``	``	``	8	``	28 00
W4049.	``	``	``	``	``	``	10	``	32 00

WILLYOUNG STANDARD MICA CONDENSERS.

(Guaranteed accurate to $\frac{1}{4}$ %).

Mounted in polished mahogany or cherry cases with polished hard rubber tops.

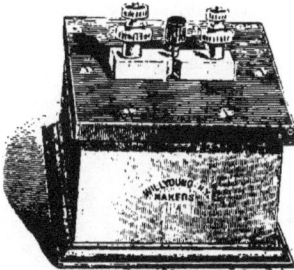

W4055. **1-3 Micro-farad** { Grade A $60 00 / " B 45 00 }

W4056. **1-2 Micro-farad** { Grade A $65 00 / " B 50 00 }

W4057. **1 Micro-farad** . { Grade A $75 00 / " B 57 50 }

Fig. 5. W4055 to W4057.

In the condensers following W4058 and W4059, the condenser sections are connected between the parallel brass blocks just as are the coils in a "multiple-arc" resistance box, so that they may be joined either in series or multiple or in combination of series and multiple, thus giving a much greater range of capacities for the same individual sections than can be secured by the orthodox pattern, where only multiple combinations are provided.

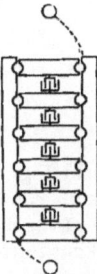

Fig. 6. Plan of W4058.   Fig. 7. W4058.

4058. **1 Micro-farad** ....Grade A $100 00
                                              " B 70 00

Subdivided into 5 sections of 0.05, 0.05, 0.2, 0.2 and 0.5 M. F's.

4059. **1 Micro-farad** ....Grade A $200 00
                                              " B 140 00

Subdivided into 12 sections of 0.001, 0.002, 0.002, 0.005, 0.01, 0.02, 0 02, 0.05, 0.1, 0.2, 0.2 and 0.5 M. F's.

WILLYOUNG " D. P." STANDARD MICA CONDENSERS.

The "D. P." (Decimal Plan) Standard Mica Condensers are the device of Mr. Willyoung and Dr. E. F. Northrup. With them a greater number of values may be obtained from a limited number of sections than with any other plan of condenser heretofore made. These obtainable values are related to one another in a strictly logical, simple and easily manipulated decimal plan. As contrasted with other arrangements for obtaining a large number of values from a small number of sections the "D. P." plan is immensely superior in that *any desired value* can be *instantly secured with no calculation*, while the corresponding value for any disposition of the plugs is also equally readable without calculation, whereas in other plans the calculation necessary to sum up the several combinations of series multiple, etc., becomes very complex and confusing.

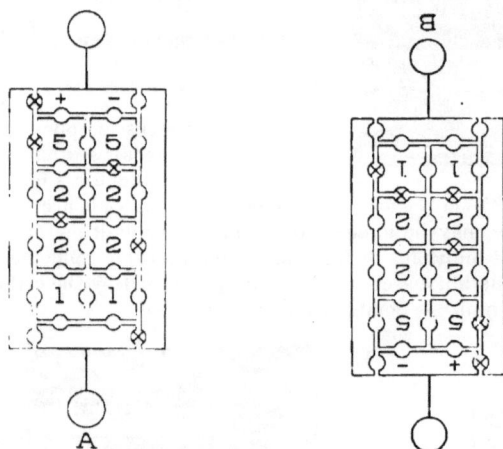

Fig. 8.

The arrangement of the "D. P." condensers is shown in Figs. 8 and 10, A, B, C and D. A and B represent each a single bank condenser, while C and D represent a two-bank condenser. As many of these banks may be combined together as is desired. Each bank is laid out in the 1, 2, 2, 5 plan. familiar in resistance boxes ; the individual sections being joined between parallel pairs of blocks as shown in the diagrams. Bars at ends and sides permit the various sections to be multipled in any combination desired, while the other sections are left absolutely out, or, by using the traveling plugs each bank may be split up so as to make two other different condensers at the same time, which may then be used on as many different circuits. Besides parallel combinations, the sections may be joined in series and in combinations of series and parallel. Each section also may be separately short circuited. Different banks also may be joined in series or in parallel with one another.

The simplest use of the " D. P." Condenser is by simple multiple connection of its sections. Thus, if we have a single bank condenser, with the four sections respectively, 0.5, 0.2, 0.2 and 0.1 M. F., we may, by multiple-ing, get successively, 0.1, 0.2, 0.3, 0.4, etc., to and including 1.0 M. F. —in other words, we may have such a condenser instantly set to the tenths' place. If now we have a two-bank condenser, one bank as just cited and the other in hundredths, we may now set to tenths with the first bank and to hundredths with the second. In other words, we may break the total of the two-bank condensers into one hundred equal parts and instantly set off any one of the hundred different values. With a three bank condenser we break into 1,000 equal divisions, etc.

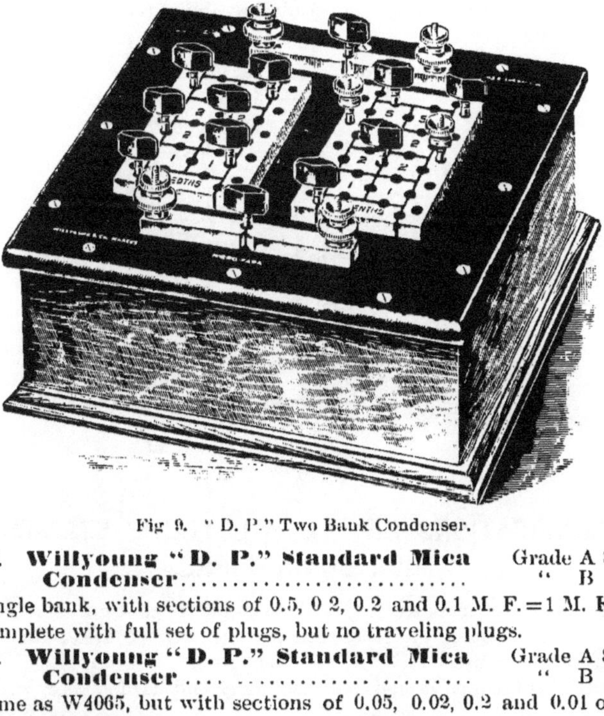

Fig 9. " D. P." Two Bank Condenser.

W4065. **Willyoung " D. P." Standard Mica**   Grade A $115 00
       Condenser ....   "   B   75 00
Single bank, with sections of 0.5, 0 2, 0.2 and 0.1 M. F. = 1 M. F. total. Complete with full set of plugs, but no traveling plugs.

W4066. **Willyoung " D. P." Standard Mica**   Grade A $110 00
       Condenser ....   "   B   70 00
Same as W4065, but with sections of 0.05, 0.02, 0.2 and 0.01 of M. F. = 0.1 M. F. Total.

W4067. **Willyoung " P. D." Standard Mica**   Grade A $110 00
       Condenser ....   "   B   70 00
Same as W4065, but with sections of 0.005, 0.002, 0.002 and 0.001 M. F. = 0.01 M. F. total.

W4068. **Willyoung " D. P." Standard Mica**   Grade A $200 00
       Condenser ....   "   B   130 00
Two banks, with sections of 0.5, 0.2, 0.2 and 0.1 in first bank ; sections of 0.05, 0.02, 0.02 and 0 01 in second bank = 1.10 M. F's total. With full set of plugs, but no traveling plugs.

W4069. **Willyoung " D. P." Standard Mica**   Grade A $205 00
       Condenser. ....   "   B   135 00

Two banks, with sections of 0.05, 0.02, 0.02 and 0.01 in first bank ; sections 0.005, 0.002, 0.002 and 0.001 in second bank = 0.11 M. F. total. With full set of plugs, but no traveling plugs.

W4070. **Willyoung " D. P." Standard Mica**   Grade A $250 00
       Condenser ....   "   B   160 00

Three banks, with sections of 0.5, 0.2, 0.2 and 0.1 in first bank ; sections of 0.05, 0.02, 0.02 and 0.01 in second bank : and sections of 0.005, 0.002, 0.002 and 0.001 in third bank = 1.11 M. F. total. With full set of plugs, but no traveling plugs.

W4071. **Traveling Plugs** for isolating individual sections, per pair $1 00

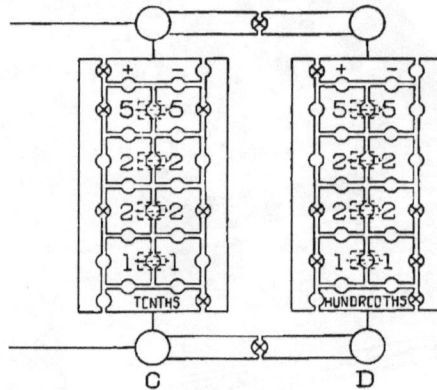

Fig. 10.

SPECIFICATIONS OF ALL WILLYOUNG STANDARD MICA CONDENSERS.

Guaranteed accurate to $\frac{1}{4}$ %, and to stand a continuous alternating E. M. F. (usual frequency) of 250 volts without risk of break-down or undue heating ; or a direct E. M. F. of 1,000 volts.

Grade A is finished in polished mahogany and hard rubber, with brass work finely polished and lacquered.

Grade B is finished in medium polished cherry, with polished hard rubber top. Brass work, a neat working finish.

We guarantee Grade B to have fully as good working qualities, and to be fully as substantial and accurate in every way as Grade A.

————— - —————

Mica Condensers guaranteed to stand 500 volts continuous alternating E. M. F.; or 1,500 volts direct E. M. F. will be made in any of the listed designs at an additional charge.

AIR CONDENSERS.

W4075. **Air Condenser,** Ayrton & Perry's Pattern..... ......$75 00

Of glass and tin foil ; may be dissected and the capacity varied at will. (*Ayrton. Fig. 129, p. 335, 3d Ed*)

W4076. **Standard Air Condenser** ....*Price on application.*

As used by the British Assoc. Elec'l Stnds. Committee. Of concentric brass tubes supported on insulating material. C=about 0.005 M. F.

W4077. **Standard Air Condenser, Kelvin's Spherical Form.**
Price on application.

(See *Gray, Vol. 1, p. 420*) Newly designed and improved.

WILLYOUNG HIGH POTENTIAL CONDENSER.

For Experiments with Currents of High Potential and Frequency.

These condensers are constructed of sheet glass and thin copper sheet. After the condenser has been stacked up, it is placed in a melted wax compound, and the air exhausted while the wax is cooling. By this process the condensers are given the perfect insulation ordinarily obtained with oil.

They are guaranteed to withstand a three quarter inch discharge (produced by a high potential transformer, Holz, or other form of static machine), across a parallel discharge gap. The wax insulation prevents all surface leakage, and no brush discharges are visible.

The glass plates of these condensers measure 13"x9" in size ; in the completed state the condensers run about 2" thick to the $\frac{1}{100}$ M. F.

PRICE LIST.

W4080. **Willyoung High Potential Condenser**.........$30 00

Described above—0.01 M. F. capacity, and guaranteed to stand a ¾" spark. Mounted in nicely-finished cherry-box, with well-insulated terminals.

W4081. Same as W4080, but 0.02 M. F. capacity $42 00

W4082. Same as W4080, but 0.05 M. F. capacity.................. 60 00

W4083. **Willyoung Adjustable High Potential Condenser**........... .... 85 00

With total capacity of 0.05 M. F. divided into individual sections of 0.005, 0.01, 0.01, and 0.025 M. F. The various arrangements of capacity are effected by sliding connecting pieces attached to the top of the containing case.

W4084. Same as W4083, but $\frac{1}{10}$ M. F. in all, divided into ten sections of $\frac{1}{100}$ M. F. each............................. ....$125 00

ELECTROMETERS.

An Electrometer is essentially a condenser in which one conducting surface (or set of conducting surfaces) is arranged to move in some uniform manner with reference to the other conducting surface when the two are subjected to the electrostatic stresses of opposite charges. The motion of the movable number may be against gravity, as in the Kelvin Absolute Instruments, against torsion as in the various Quadrant Electrometers, or against any other style of control which is thinkable.

Law of the Electrometer – Suppose the two conducting elements to be charged, respectively, to potentials V_1 and V_2. Then

$$D = K \ V_1 \ V_2 \ \ldots\ldots\ldots\ldots\ldots(4)$$

or *The deflections are proportional to* THE PRODUCT OF THE CHARGES.

Suppose a battery or other generator is joined to the Electrometer Plates. Then $V_1 = V_2$, and

$$D = K \ V^2 . \ \ldots\ldots\ldots\ldots\ldots(5)$$

i. e.--*The deflections are proportional to the* SQUARE *of the E. M. F.*

Suppose one plate to be joined to a point of fixed potential and the other plate to any unknown potential to be measured. Then V_1 – a constant and $K \ V_1 = K = $ a new constant so that

$$D = KV \ \ldots\ldots\ldots\ldots\ldots(6)$$

i. e.--*The deflections are directly proportional to the Potential to be measured.*

Uses of Electrometers -- The law (5) applies equally well to alternating as to direct E. M. F's, and is independent of the frequency since *both* plates change the sign of their charge *simultaneously :* hence, the sign of their *product* does not change. The instrument is thus very useful for accurate E. M. F. measurements on A. C.'s, and where the E. M. F.'s are small, say less than 10 volts, is almost the *only* reliable method. Alternating *currents* may be measured by passing them through a non-inductive resistance and taking the drop with the electrometer.

In the measurement of the specific resistance of insulators, and of the I. R. of very short lengths of high insulation cable by the "leakage method," the electrometer is invaluable.

W4100. **Modified Mascart Electrometer**.... ...Grade B $80 00

With exceedingly high insulation, accessible needle and facilities for drying out with hot air.

W4101. **Ryan Electrometer**......................Grade B $80 00

For obtaining E. M. F. and current curves of alternating machines and transformers. A coil wound upon the instrument enables the electrostatic couple to be balanced by an electro-magnetic one, so as to use the electrometer as a zero instrument.

W4102. **Carhart's Electrometer or Static Voltmeter**...

Grade B $125 00

(See *Proc. of the International Electrical Congress*, 1893, p. 208 ; also *E. W.*, Sept. 16, 1893 ; also C. & P., pp. 200–204). For obtaining E. M. F. and current curves of alternating machines and transformers. Indicates up to 1100 volts.

W4103. **Carhart's Electrometer or Static Voltmeter.**

Grade B $100 00

Same as W4102 but smaller ; indicating up to 110 volts.

W4104. **Simple Quadrant Electrometer.** For School and Students' use. All parts are accessible and the insulation is good. Will deflect for one cell of battery.......................................$20 00

GALVANOMETERS.

Types of Galvanometers—Broadly speaking there are but two types of galvanometer in use :

    a. Coil fixed and magnet moving as *e. g.*, the KELVIN type.

    b. Magnet fixed and coil moving as *e. g.*, the D'ARSONVAL type.

The Kelvin Galvanometer — Also spoken of as the THOMSON galvanometer. By this is generally understood the reflecting mirror galvanometer of either two or four coils ; the magnetic system generally consists of a number of short steel needles attached to the back of the reflecting mirror and the whole is suspended by a fine torsion fibre of cocoon or quartz. The system may or may not be astatic.

The D'Arsonval Galvanometer has a coil suspended in the field of a magnet, and with its axis at right angles to the field, by a metallic wire or fibre ; there is a similar fibre below either straight or coiled. Current is passed through the coil by use of these fibres and deflection of the coil takes place.

Comparative Merits and Demerits—Sensitiveness: The Kelvin type is the most sensitive, the average working sensitiveness being from 30,000 to 60,000 megohms, while in specially delicate work sensitivenesses up to millions of megohms have been obtained. The corresponding figures for the D'Arsonval are 800 to 1,000 megohms for an average working sensitiveness and 2,000 to 2,500 megohms for the best maximum.

---

* The various ABSOLUTE, QUADRANT and PORTABLE Electrometers of the Kelvin type as made by James White, Elliott Bros., etc., will be quoted on and particulars furnished on application.

Definition — Sensitiveness : *The Sensitiveness of a galvanometer is a measure of the ratio of deflection produced to current producing it.*

In American Instrument practice the term is generally used to designate the *resistance* which the circuit must have in order that one volt applied to its terminals may produce exactly one division deflection.

Dead-Beatness—The D'Arsonval is naturally a very dead beat instrument since, when the coil deflects, eddy currents are set up which tend to bring it to rest. Sometimes the coil is wound on a little frame of copper or aluminum foil in order to increase this effect.

The Kelvin galvanometer is not "dead-beat," except when the mirror or (sometimes) a mica or aluminum vane attached to the system is enclosed in an air box.

Other Galvanometers. Both the Kelvin and D'Arsonval forms of galvanometer are found in various forms depending upon the special duties they may have to perform. Where the instrument need not be very sensitive the suspensions are replaced by springs or, if of the Kelvin type, the system is pivoted in jewels, the earth's magnetic field supplying the controlling force.

Choice of a Galvanometer. The D'Arsonval galvanometer is always the most convenient since it is "dead-beat," not affected by neighboring magnetic fields, and has deflections (generally) strictly proportional to the current flow. It is also not seriously disturbed by mechanical jar, making it, hence, very suitable for factory and marine work. Its sensitiveness is amply great enough for almost any purpose except certain spectroscopic and purely scientific research work.

For the measurement of insulation resistance the D'Arsonval galvanometer has replaced the Kelvin form very generally, and is giving most satisfactory service.

The Ballistic Galvanometer. This may be of either the D'Arsonval or Kelvin type and is distinguished only by its having a very heavy and practically "undamped" system. Its time of vibration is generally long. It is employed in the measurement of transient currents as *e. g.* induction currents in hysteresis tests, the instantaneous currents of condenser discharges, etc.

Law of the Ballistic Galvanometer. With a heavy system and a transient current, the latter has ceased to flow before the system has had time to start off. Theory shows (S. & G., Vol. II, page 363) that in this case *the current is proportional to the first swing expressed in scale divisions.*

The Differential Galvanometer is distinguished by its having two separate windings *of the same resistance* and each producing the *same permanent deflection* when the same voltage is applied to the terminals. In many special measurements it is very useful.

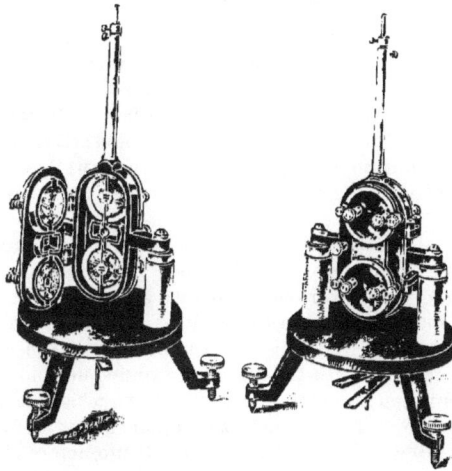

Fig. 11.   W4110, etc.

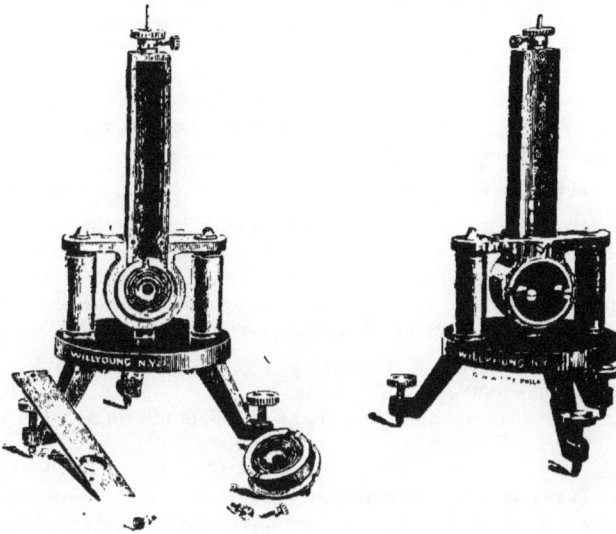

Fig. 12.   W4125.

KELVIN REFLECTING GALVANOMETERS.

The following Kelvin Galvanometers are provided with coils of but 1¼" diameter by ⅜" thick while the magnetic needles making up the system are but a few mms long. The mirrors are very small and thin but very perfect; the entire system thus weighs but a few milligrams. Both theory and practice during the past few years have conclusively shown these conditions to secure enormously higher sensibilities as well as greater deadbeatness and speed of working than are given by Kelvin Galvanometers as previously made.

Coils of the size of above are not over half the diameter and thickness of the coils heretofore used in "Thomson" Galvanometers. The cubic volume is therefore not over one-eighth as great as in the old fashioned coil. Consequently the resistance will be, for the same gauge of wire, only about one-eighth as much. Nevertheless the sensitiveness will be many times as great as with the higher resistance but old style instrument.

Our 2000 ohm galvanometer is much more sensitive than the 10,000 ohm instrument heretofore manufactured and still offered by most makers.

We do not advise the use of four coil galvanometers of more than 5000 ohms resistance, or two coil instruments of more than 2500 ohms, as these instruments can be made enormously sensitive, and the wire used for their winding (No. 40 B. S.—0.0031 inches diameter) is as fine as can be conveniently handled. For a 10,000 ohm galvanometer (four coil) 0 0024 wire must be employed.

In all galvanometers of the Kelvin type, whether two coil or four coil, ballistic or ordinary, the coils are enclosed in standard interchangeable coil frames or boxes; *these coils, may, therefore, be used at any time in any one of our galvanometers.*

In all of our Kelvin Galvanometers the coil frame supports are suspended above the base by corrugated hard rubber pillars; a brass tube telescopes over these pillars to protect them from dust when not in use while affording the highest attainable insulation when in use.

W4110. **Kelvin Galvanometer** (Four Coil)......... Grade B $65 00

Improved "Scissors" control magnet under base of instrument. Resistance 5000 ohms. Four coils so arranged that they can easily be joined in any combination of series or multiple. Front is hinged, to swing open and expose the needle. Axis of rotation lies in plane of mirror. Coils and frames are all made to standard jigs and are interchangeable.

W4111.	Similar to W4110; 100 ohms.............. ....Grade B $55 00
W4112.	Similar to W4110; 1000 ohms...................	"	B	60 00
W4113.	Set of 4 Interchangeable coils only; 5000 ohms, total	"	B	35 00
W4114.	"	"	" "	100	"	"	" B 25 00
W4115.	"	"	" "	1000	"	"	" B 30 00

W4116. **Kelvin Galvanometer** (Four Coil).........Grade B 70 00

Similar to W4110, but with additional arrangement for swinging needle out from the coils for purposes of inspection, freeing from lint, etc. Resistance 5000 ohms.

W4117   Similar to W4116; 100 ohms.....................Grade B $60 00
W4118.   "   W4116; 1000 ohms.......... ....   " B 65 00
    Sets of Interchangeable coils: same prices as W4113 to W4115.
W4119 **Kelvin Galvanometer** (Two Coil)...........Grade B $50 00
    Designed to take the place of the "tripod pattern," to which it is vastly superior. "Scissors" control magnet beneath base of instrument. 5000 ohms resistance.
    The general construction of this instrument is similar to that of W4125.
W4120.   Similar to W4119; 100 ohms...... ....Grade B $40 00
W4121.   "   W4119: 1000 ohms............. ......   " B 47 00
W4122.   Interchangeable coils only; 5000 ohms...........   " B 27 50
W4123.   "   "   100   "   ....   " B 20 00
W4124.   "   "   1000 "   ....   " B 22 50
W4125. **Ballistic Galvanometer** ....   " A 80 00
                                               " B 65 00

Improved form, having two small coils which may be combined in series or multiple. Has improved adjusting and controlling magnet and high insulation. Period and sensitiveness may be varied within wide limits. Torsion head divided on silver. Coils and entire system are accessible without removing case. Resistance 500 ohms.

WILLYOUNG HIGH SENSIBILITY D'ARSONVAL GALVANOMETER.

    Our High Sensibility D'Arsonval Galvanometer is of recent design and combines the best features of earlier instruments with a number of improvements.

    The general construction of the instrument is evident from the illustration. The magnets are thin and of the best special magnet steel; they are hardened and magnetized by an improved process tending to secure maximum magnetization and permanence of the same.

    Both Dead Beat and Ballistic Systems may be used with this instrument. The systems are mounted in tubes which may be removed from the instrument very quickly, to make room for others; in making this change the connections are automatically made and broken. A clamp upon the tube holds the coil firmly and takes all strain from the suspension when the system is not in use. In case of accidental breakdown of the suspension the tube of *the coil system may be completely removed from the inner rib sustaining the system itself : the latter is thus extremely accessible.*

    The coils are made from purest obtainable copper insulated with pure white silk. The suspensions are of artificially aged phosphor bronze strip of which the "set" is inappreciable.

    Owing to the shape of the pole pieces *deflections are proportional to the deflecting currents.*

    The moment of inertia of the moving system is so adjusted that the instrument is very little affected by mechanical vibration. *It is entirely undisturbed by neighboring electro magnetic changes.* These features combined with the exceedingly high sensibilities which we are able to give to these instruments make them greatly superior to Kelvin Galvanometers for the majority of purposes. For cable and other high insulation testing the Willyoung High Sensibility D'Arsonval Galvanometer is most highly recommended.

As purchasers may desire several different coils we list the Galvanometer and various tubes separately.

W4135. **Willyoung High Sensibility D'Arsonval Galvanometer.** $40 00
Complete except tube.

W4136. **Interchangeable Tube for Willyoung High Sensibility D'Arsonval Galvanometer** 13 00
Resistance about 20 ohms.   Sensibility about 80 megohms.

W4137. **Tube for Willyoung High Sensibility D'Arsonval Galvanometer** 14 00
Resistance about 200 ohms.   Sensibility about 200 ohms.

W4138. **Tube for Willyoung High Sensibility D'Arsonval Galvanometer** 15 00
Resistance about 1000 ohms.   Sensibility about 800 megohms.

W4139. **Tube for Willyoung High Sensibility D'Arsonval Galvanometer** 20 00
Resistance about 3500 ohms.   Sensitiveness about 1500 megohms.

If ordered from list above all tubes will be furnished *"deadbeat" and with plane mirrors.* Tubes ballistic, and with either plane or concave mirrors will be furnished at the same prices, if so ordered.

These galvanometers are made exactly the same in construction as the telescope and scale forms just described, and possess exactly the same features.

THE "AONE"
COMBINATION WALL AND TABLE D'ARSONVAL GALVANOMETER

is the best all round galvanometer yet produced. Is complete with telescope and scale, and changes instantly from table to wall and vice versa. Always ready, very portable, and cannot get out of order. All parts interchangeable.

See special descriptive pamphlet published separately and furnished on application.

Fig. 13. W4135, etc.

WILLYOUNG H. S. GALVANOMETER.

TRIPOD D'ARSONVALS AND OTHER GALVANOMETERS.

W4155. Tripod D'Arsonval Galvanometer.......... ... $20 00

Table form, with interconvertible damped and undamped coils. Resistance of coil about 1,200 ohms. Sensitiveness (with coil damped) about 200 megohms.

W4156. Tripod D'Arsonval Galvanometer............ $24 00

Same as W4155, but with low resistance coil for thermal work.

W4157. Coil only of W4155, with damper and mirror 5 00

W4158.   "   " W4156, with damper, mirror and special suspension.. 9 00

W4159. Carpentier D'Arsonval Galvanometer $65 00

This is the original form of the D'Arsonval instrument as made by Carpentier of Paris. The magnet is vertical, and a cylindrical iron core is placed between the poles. The instrument is handsomely finished in polished hard rubber and lacquered brass and enclosed in glass; it presents a handsome appearance.

W4160. Simple Tangent Galvanometer................$12 50

Frame of well dried cherry; main diameter of coil about $7\frac{3}{8}$" with three windings of 1, 5 and 10 turns, arranged for use separately or in series. Glass pointer and 3" diameter degree scale. System is pivoted in a glass jewel and cannot be displaced for any position of the instrument.

W4162. Universal Tangent Galvanometer....... ...... $45 00

May be used as Tangent, Gaugain, Hemholtz-Gaugain, Sine, Wiedemann and Detector Galvanometer. Measures currents ranging from .000002 amp. to 100 amperes (see Carhart's *Electrical Measurements*). Coil frames of well polished cherry "built up" to prevent warping. *Each instrument furnished with a certificate of all dimensions not obtainable from the finished instrument.*

Fig. 14. W4155-56. TRIPOD D'ARSONAL.

Fig. 15. W1152.

Fig. 16. W4163.

W4163. **Pocket Detector Galvanometer**........... .... $4 00

Handsomely finished in hard rubber mounting. A hard rubber cap slips over the instrument and completely encloses it when not in use. Makes a very convenient pocket *voltmeter* for storage battery testing. Resistance, about ½ ohm.

Fig. 17. W4164.

Fig. 18. W4165.

W4164. **Horizontal Detector Galvanometer**............$4 00

W4165. **Vertical Detector Galvanometer** 10 00

PORTABLE TEST D'ARSONVALS.

This instrument is built along the lines of the galvanometer used in our "AONE" Testing Sets. It has a scale about 2" in length, divided into 50 divisions, has deflections proportional to currents, and is exceedingly sensitive, so that it may be used to great advantage in insulation measurements by direct deflection methods. It is mounted in a polished mahogony case with lid, lock, and handle.

W4164.   **P. T. D'Arsonval;**.............................. . $30 00

Resistance 400-500 ohms.   Sensitiveness about one megohm.   Measures about 5"x3"x3¼" deep over all

W4165.   **P. T. D'Arsonval and Battery**...................$40 00

Same as W4164, but with six cells of Silver Chloride Battery, mounted in case so that any one or more cells may be used at will.   Measures outside about 5"x4"x3¼" deep over all.

W4166.   **"Aone" Portable D'Arsonval***.................$100 00

For cable insulation; testing, both indoors and out, and for all classes of work requiring a portable, sensitive, convenient and ever ready instrument.

This instrument is an improved form of the Portable Galvanometer (B5355) lately made by Willyoung & Company of Philadelphia.   The galvanometer is a D'Arsonval hung on gimbal bearings with devices for quick leveling.   The front of the instrument is held by a bayonet catch, thus being instantly removable and making the entire interior accessible.   A clamp controlled by a screw at the left of the suspension tube raises the weight of the coil from its suspension and holds it firmly when not in use. The telescope and scale are attached to the instrument proper so as to be always in adjustment.   The insulation of the instrument has been greatly improved and is now very perfect.   The coil is one of our improved convertible type and while supplied as a "dead-beat" coil is instantly changed to a ballistic coil, by simple removal of a sliding aluminum air damper attached to the system.

When not in function the galvanometer turns down flat and locks fast to the telescope and telescope support while the scale slips out and inserts into a pair of clips arranged to receive it.

* Complete in polished oak carrying case; measures 22½x10½x7" deep over all and weighs about 18 pounds   Sensitiveness about 750 megohms.

W4167.   **Portable H. S. D'Arsonval***....................$110 00

For out and indoor insulation testing, when the very highest sensitiveness is required.   Complete with telescope and scale.   This is our galvanometer W4135, etc., but arranged for the maximum of convenience and portability.   The galvanometer is supported in a ball and socket joint and levels by use of three screws without detaching from its case.

In polished oak carrying case with lock, handles and carrying strap. Sensitiveness, etc., as per W4138.

W4168.   **Portable H. S. D'Arsonval**....................$115 00

Same as W4167, but sensitiveness, etc., as per W4139.

___

Notes.—For tripod and stool for street use with this and similar instruments, see index, "Tripods."

GALVANOMETER MIRRORS.

Of silvered glass and very accurately plane Warranted free from distortion and optically perfect.

W4180.	**Concave Mirror.**	$\frac{1}{4}$ in. diam., 36 to 42 in. radius			....$1 50
W4181.		$\frac{3}{8}$	"	"	"	.... 1 00
W4182.		$\frac{1}{2}$	"	"	"	... 1 25
W4183.		$\frac{3}{4}$	"	"	"	.... 1 50
W4184.	**Plane Mirror.**	$\frac{1}{4}$ in. diam., 1-50 to 1-20 in. thick			.... 1 50
W4185.		$\frac{3}{8}$	"	"	"	.. ... 1 25
W4186.		$\frac{1}{2}$	"	"	"	..... 1 75
W4187.		$\frac{5}{8}$	"	"	"	...... 2 00
W4188.		$\frac{3}{4}$	"	"	"	...... 2 00
W4189		1	'	"	"	...... 3 25

Steel mirrors of any diameter and thickness, and polished on one or both sides will be furnished to order, and prices furnished on application. *We guarantee the finest attainable results in this line.*

CIRCULAR SCALES.

W4190. **Circular** (3" diam.) **Scale** ... ...each $0 10

On bristol board from engraved plate ; graduated in degrees and numbered from zero to 90° each way from both ends of the same diameter.

Per dozen, 1 00

W4191. **Circular** (4" diam.) **Scale** ....each   15
"   4"   '   "   ....per dozen 1 50

Same as W4191 except for the larger diameter.

GALVANOMETER SUSPENSIONS.

D'Arsonval Suspensions. The suspensions for any given style of our D'Arsonval Galvanometers are always made of exactly the same dimensions and are tipped with No. 22 copper wire to fit our standard suspension rod. They are thus readily fitted and solid connection made without the use of solder.

W4192. **Upper Suspension** (Straight).. .. .. .each $ .25
"   "   "   ....per dozen 2.50

W4193. **Lower Suspension,** (Coiled)......... ....each   .50
"   "   "   ....per dozen 5.00

*Suspension Strip** for the above suspensions; rolled from $1\frac{1}{2}$
mil. or 2 mil. (as ordered) phosphor bronze wire; thickness
about ½ the width ....per foot   .15

Per spool of 10 feet 1.50

W4194. **Cocoon-fibre** for Kelvin Galvanometers; reel of about
20 feet.. .10

* Unless specifically stated otherwise, all our D'Arsonval Suspensions are of the $1\frac{1}{2}$ mil. strip.

Quartz Fibres. Quartz may be fused and drawn into fibres of extreme fineness. In this form quartz makes a most perfect suspension for galvanometer and similar systems. The chief points of merit are:

A—Great tensile strength—sixty to seventy tons per square inch of cross section: thus nearly equal to the best steel.

B—Perfect Torsion—*i. e.* the twist exactly proportional to the moment of turning force.

C—Perfect Elasticity—hence absolute return to zero.

D—Perfect insulating properties whether in moist or dry atmospheres.

CONSTANTS OF QUARTZ FIBRES.

(Threlfall, p. 218, etc.)

Young's Modulus @ 20° C....................5.6x10' C. G. S.

Modulus of Simple Rigidity @ 20° C.... ....2.65x10'' C. G. S.

Modulus of Incompressibility ....1.4x10'' C. G. S.

Modulus of Torsion.................... ......3.7x10'' C. G. S.

Temp. Coef. of Modulus of Torsion, 22° C to 98° C... .000133

Coefficient of Linear Expansion (approx.) .... .0000017
    Between 30° C and 80° C.

Limit of allowable twist .... .⅓ turn per cm for fibre .01 cm diameter

Diameter of fibre breaking with 10 grammes... .. .0014 cm or .0007 inch

The fibres listed below will carry from 1 to 30 grains and vary from .0002 to .00015 in. in diameter. Directions for manipulating the fibres accompany each box.

W4195. **Quartz-fibres.** Set of 6 in box; each 6 in. long........$3 00

W4196. **Quartz-fibres.** " " " 12 " 4 00

W4197. **Quartz-fibres.** " " " 24 " 6 00

GALVANOMETER SHUNTS.

These are employed to reduce the sensitiveness of a galvanometer at will. They are of wire, preferably of the same material as the Galvanometer coils so as to vary resistance with temperature in the same proportion, and mounted in a suitable case. Fig. 21 shows in diagram a galvanometer with three shunts for sending respectively $\frac{1}{10}$, $\frac{1}{100}$ or $\frac{1}{1000}$ of the current through the galvanometer according to the location of the plug. The actual resistances are $\frac{1}{9}$, $\frac{1}{99}$ and $\frac{1}{999}$ of the galvanometer resistance. Such a box of shunts is, of course, suited to the given galvanometer only.

"**Multiplying Power**" is the figure denoting the ratio of total current in the combined galvanometer and shunt circuit to current actually flowing through the galvanometer.

Compensated Shunts. In such a shunt as just described the total resistance (galvanometer and shunt combined in parallel) varies according to the particular shunt in use. In many cases this circuit must be kept constant. Extra resistances are then placed within the shunt box so that simultaneously with the insertion of a lower resistance shunt, sufficient *extra* resistance is placed in circuit to bring the total up to the old figure. Such a shunt is known as a **Compensated Shunt.**

The Universal Shunt; often called the "Ayrton" shunt is shown schematically in fig. 20. It can be shown mathematically (see Proc. IEE., (*Lon.*), 1894; Elec'n, (*Lon.*), Vol. XXXII, p. 627; Elec'l World, April 21, 1894, p. 541 — *ibid*, May 12, 1894, p. 648 — *ibid*, May 19, 1895, p. 681) that the ratios of current flowing through the galvanometer are always correct *whatever the actual resistance of the galvanometer.* The same shunt will answer, therefore, for any galvanometer. With D'Arsonval Galvanometers where the suspension makes up an appreciable part of the resistance this is a very important feature because the replacing of broken suspensions usually means an altered resistance and, therefore, inaccurate shunt ratios when the old form of shunt is employed.

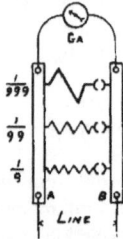
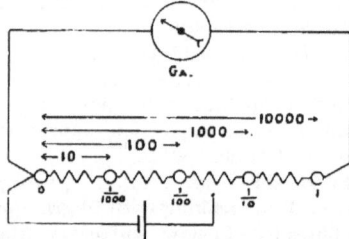

Fig. 19.   Fig. 20.

Furthermore, the **Universal Shunt is accurate for ballistic work**—not true of the old form.

It should be noted that with this shunt the galvanometer is *always* "short circuited" by the total of the shunt. The initial sensitiveness of the instrument with a 1:1 ratio is less, therefore, than would be that of the instrument itself. This decrease of initial sensitiveness is, however, very small provided the total shunt resistance is relatively large as compared with that of the galvanometer. Three ranges of this shunt are, therefore, listed.

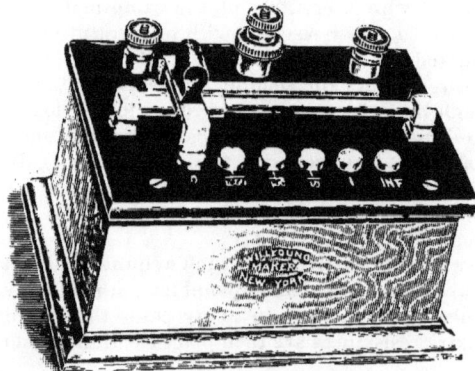

Fig. 21. W4200 to W4202.

W4200. **Ayrton Universal Shunt**................Grade B $40 00

Of about 100000 ohms; for Galvanometers having resistance of 20000 ohms or upward.

W4201. **Ayrton Universal Shunt**........ ....Grade B $27 50

Of about 10000 ohms; for galvanometers of from 2500 to 6000 ohms resistance.

W4202. **Ayrton Universal Shunt**................Grade B $22 50

For galvanometers of 2500 ohms or less.

Ayrton Shunts of Grade A will be made to order at an advance in price of $8.00.

We also make the above Ayrton Shunts with three extra ratios of $\frac{1}{2}$, $\frac{1}{20}$ and $\frac{1}{200}$, *i. e.*, the steps are $\frac{1}{1000}$, $\frac{5}{1000}$, $\frac{10}{1000}$, $\frac{50}{1000}$, $\frac{100}{1000}$ and $\frac{500}{1000}$. This makes it always possible to get a good sized scale deflection.

W4200A. **"Six Step" Ayrton Shunt**......................$50 00

W4201A.   "   "   .. 37 50

W4202A.   "   "   32 50

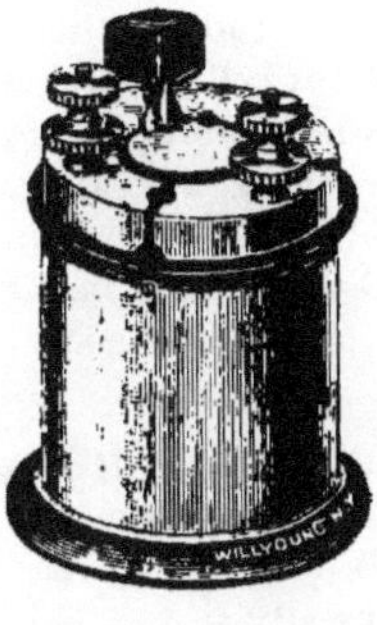

Fig. 22. W4205.

W4205. **Shunt Box** ....$\left\{\begin{array}{l}\text{Grade A \$30 00} \\ \text{ "   B 25 00}\end{array}\right.$

Orthodox form for galvanometers of 10000 ohms resistance or less; reduces galvanometer current to 1-10, 1-100, and 1-1000 of full value. Plug contacts, etc. Must be adjusted for a particular instrument.

W4206. **Compensated Shunt Box**..............$\left\{\begin{array}{l}\text{Grade A \$60 00} \\ \text{ "   B 50 00}\end{array}\right.$

Maintaining resistance of galvanometer circuit constant regardless of the shunt value used. For galvanometers of less than 10000 ohms resistance.

W4207. **High Insulation Shunt Box**..........$\left\{\begin{array}{l}\text{Grade A \$45 00} \\ \text{ "   B 37 50}\end{array}\right.$

For cases where the very highest insulation is desired, as in testing high insulations, etc. Box and Segments are hung from tall corrugated rubber pillars, and shunt plug is operated by a long insulating handle.

THE TELESCOPE, OR LAMP, AND SCALE METHOD.

In the reflecting type of galvanometer, electromometer, etc., we are unable to directly observe the angles through which the moving system turns owing to the latter's, usually, extremely small dimensions. We, therefore, observe the deflection of a beam of light over a graduated scale, the beam being thrown by a suitable fixed source upon a mirror attached to the moving system of the instrument, and thence reflected back upon the scale. Two arrangements are commonly employed for this purpose.

The Lamp and Scale. This method is largely used by English workers, and, in this country, by the older telegraph and cable engineers. The scale (usually either 50 cms in mms or 18″ in 40ths) is attached to a suitable stand or support, is in a plane parallel with the plane of the mirror and has its center intersected by the plane normal to the mirror. If the mirror is concave of one meter radius, then a beam of light proceeding from an illuminated vertical slit will be focused upon the scale, and this image will move over the scale in consonance with the movement of the system : a reasonably dark room is, of course, demanded. For the illuminated slit may be substituted a vertical wire stretched across a suitable convex lens. Upon the scale will then be a black line on a bright field; the room need not, therefore, be darkened.

The Telescope and Scale. If in place of the slit we place a telescope and the eye, then the divisions of the scale will move before the eye as the system deflects. A fine "cross hair" in the focus of the eye piece serves to exactly fix the division. This arrangement is exclusively employed by the Germans and by the majority of scientific observers in this country. The lighter the room the better, while the deflection can unquestionably be much more *accurately* read with the telescope than with the lamp. We recommend the telescope in preference to the lamp in the general case.

Angles. With either the lamp or the telescope and scale, and with the scale concentric with the axis of rotation of the system, the deflections will be directly proportional to the angle through which the reflected beam moves, and this angle will be *twice* the actual angle of rotation of the system. If the scale is straight and parallel with the stem of the mirror (when the system is at rest) then the scale readings are proportional to the *tangent* of the angle moved through by the reflected beam or, what is the same thing, to the tangent of twice the angle of rotation of the system.

When the work does not demand the very highest accuracy, the tangent of twice the angle may be assumed the same as twice the tangent of the angle and hence, for not too large angles, the same as the angle itself.

LAMPS AND SCALES.

We have devised a form of lamp and scale which is very substantial, convenient to use, and reasonable in price. A sharp black line is thrown in the middle of a round disc of light upon a translucent scale. Has all adjustments for zeros for galvanometers of different heights, etc. May be used without darkening room. May be read either from in front or behind the scale.

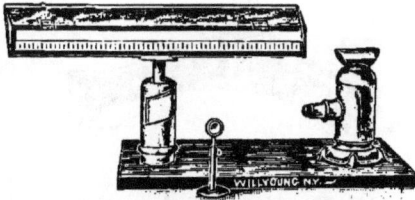

Fig. 23. W4210 to W4212.

W4210. **Improved Lamp, Stand and Scale**.....Grade B $17 00
    Using incandescent lamp.

W4211. Same as W4210, but for oil lamp................ " B 17 50

W4212. Same as W4210, but for gas burner .... " B 17 00

READING TELESCOPES AND SCALES.

These telescopes have been designed to fill the demand for an optically good instrument, simple in its adjustment, capable of giving good results even for the most exact purposes, and at a reasonable price. The clearness and definition in every case will be found very satisfactory.

W4215. **Student's Reading Telescope**................. $15 00

Has altitude and azimuth adjustments, and lateral rack and pinion adjustment of scale; aperture 15 mm; magnifying power from 12 to 15. Reads down to 60cm. from galvanometer. Of brass throughout, with lacquered finish, except base, which is japanned. A 50 cm. scale is included.

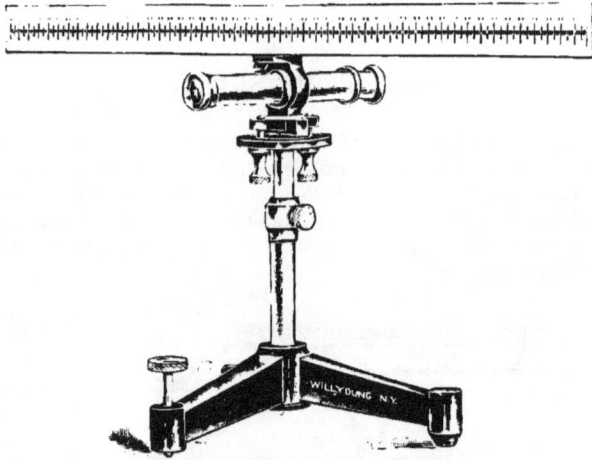

Fig. 24. W4215.

W4216. **"Universal" Reading Telescope** $25 00
Constructed of brass; lacquered throughout, except base which is japanned. Adjustments for altitude and azimuth, vertical rise and fall of scale, rack and pinion adjustment of telescope, and independent movement of eye-piece to secure distinct focus of cross hairs. Magnifying power 20, aperture 25 mm. Complete with 50 cm. scale.

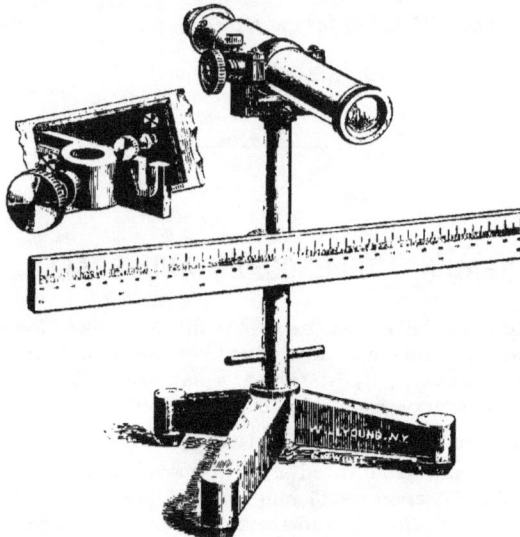

Fig. 25. W4216.

Fig. 26.   W4220.

W4220. Unmounted Reading Telescope................ $6 00

As furnished with D'Arsonval Galvanometers but without support of any kind; will focus down to 16''.

W4221. Cardboard Scale (unmounted)..................... $ 25

50 cm. long; graduated in mm, but unnumbered.

W4222. Cardboard Scale (unmounted)..................... 50

Same as W4221, but numbered.   Zero at one end.

W4223. Cardboard Scale (mounted)...................... 1 00

Same as W4221, but mounted upon a "built up" wood back.

W4224. Cardboard Scale (mounted)....................... 1 25

Same as W4222, but mounted upon a "built up" wood back.

W4225. Mounted Translucent and Opaque Scale......$14 00

As made for the New York Telephone Company.   A card board, 100 cm (mm division) scale, zero in center, one-half red and one-half black numbers, is glued upon a strip of ground glass having about twice the width of the scale itself.   The whole is then framed in polished mahogany and arranged so as to adjust vertically upon two supporting standards. The observer has the scale between himself and the galvanometer, numbers facing him, and sees the lamp image through the ground glass and against the edge of the numbered scale.

THE WHEATSTONE'S BRIDGE.

This is shown diagramatically in fig. 27, where A C, C B, A D and D B are four conductors of resistance R_1, R_2, etc., and joined together as shown across one pair of opposite joint points is a galvanometer, G, while a battery, B, is laid across the other two, A and B. It can be shown that if the four resistances A C, C B, etc., be varied until no current flows in the galvanometer circuit then

$$\frac{R_1}{R_2} = \frac{R_3}{R_4} \text{ or } \frac{R_1}{R_3} = \frac{R_2}{R_4} \quad \dots\dots\dots\dots\dots\dots(7)$$

and the Bridge is *balanced*. If in place of R_1, we have an unknown resistance, X, then we have

$$X = R_3 \frac{R_2}{R_4} \quad \dots\dots\dots\dots\dots\dots(8)$$

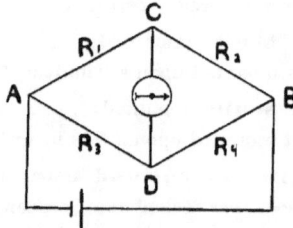

Fig. 27.

The "Wire," "Meter," or "Slide" Bridge as it is variously called, is one of the simplest forms of the Wheatstone's Bridge. The two instances R_3 and R_4 of fig. 27 are replaced by a straight and uniform wire A B whose resistance, of course, is proportional to its length. The galvanometer terminal is a form of "slider" which can make contact up in the wire anywhere in its length. When *balanced* we may write, therefore,

$$\frac{X}{R_3} = \frac{l_1}{l_2} \text{ or } X = \frac{l_1}{l_2} R_3 \quad \dots\dots\dots\dots\dots(9)$$

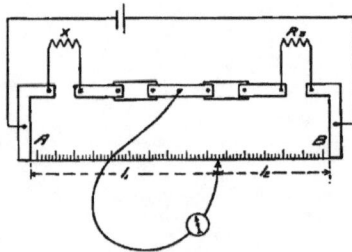

Fig. 28.

REQUIREMENTS OF A GOOD SLIDE BRIDGE.

1. The wire should be uniform in cross section and homogeneous in structure, so as to have its resistance strictly proportional to length.

2. Wire must be hard so as not to be injured by slider.

3. Scale should be divided into 100 or 1000 equal parts and its total length should be *exactly* equal to that of the wire.

4. Copper strip joining wire to "gaps" should be massive, as its resistance is assumed absolutely negligible in the measurements.

In the several Willyoung "Slide" Bridges below, the bridge wire is made for us abroad especially for the purpose; it is supplied to us in nearly straight pieces, and every effort is made to give as near perfect homogeneity throughout as is possible.

Fig. 29. W4230.

W4230. **Meter Bridge** (B A Form).......................... $12 00

With four gaps and wood scale one meter long, divided in mm's. Wire of platinoid; with improved hard rubber slider, having absolutely no play in either slider or key proper. Slider key makes automatic contact with a copper wire running parallel with the bridge wire thus avoiding the necessity of carrying the galvanometer connection dangling from the slider as it is moved along. Substantially made and neatly finished in cherry and lacquered copper.

W4231. **Meter Bridge** (Short Form).......................... $10 00

Same as W4230 but with wire only half the length, viz., 50 cms. Scale is divided in 1,000 half mm divisions.

W4232. **Meter Bridge** (B A Form).......................... $25 00

Same as W4230, but with engine divided scale on polished boxwood and numbered both ways. Copper parts are heavier than in W4230; the base also is heavier, of mahogany and cross-braced against warping. The whole instrument is highly finished.

Fig. 30. W4235.

W4235. **Improved Meter Bridge**.......................... $60 00

Scale of silvered brass one meter long, graduated in mm's, and numbered both ways. Vernier enables readings to be made to 1·20 mm. Slider is so designed that it is impossible to jam or cut the bridge wire, and has both coarse and fine adjustments. Base is of heavy mahogany, cross-braced to prevent warping, and the copper work massive and finely polished and lacquered.

This instrument has been made for such institutions as the *University of Michigan, Michigan School of Mines, U. S. Naval Academy, University of West Virginia, Syracuse University (an order of 12),* etc.

It has reversing arrangement as pictured and described in Mascart and Joubert's *Electricity and Magnetism,* Vol. II; Price's *Electrical Resistance;* Carhart's *Electrical Measurement,* § 39; it is equipped with mercury cups and binding posts. (*The cut is old and does not show the recent improvements; detailed blue print will be forwarded on request.*)

WILLYOUNG IMPROVED CAREY-FOSTER BRIDGE AND COMMUTATOR.

The Carey-Foster Method, so called, is really nothing but a slightly specialized use of the Wheatstone's Bridge. Though simplicity itself it yet constitutes the most perfect means of determining a resistance with the very greatest exactitude, and of deriving the "temperature coefficient" (the $\%$ variation of change of resistance with temperature) that has yet been suggested and it is now almost generally used for this purpose

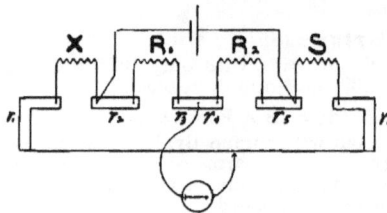

Fig. 31.

in the laboratories of the world. Consider the four gap bridge of fig. 31. Suppose X, R_1, R_2 and S to be nearly equal. If, now, we get "balance" and then exchange X and S and balance again we can readily show that

$$X = S - (x_1 - x_2) \rho \dots \dots \dots \dots (10)$$

where x_1=first balance reading, x_2=second balance reading and ρ=resistance per unit length of the bridge wire; *i. e.* the unknown, X, differs from S by exactly the resistance of bridge wire over which the slider must move in going from one "balance" to the other. This measurement will *always* be independent of the *length* of the wire and of the resistance of the various connecting strips r_1, r_2, r_3, etc.

In practice a specialized form of bridge must be employed so as to satisfy the following requirements.

(1) X and S must exchange quickly so that no change of temperature may take place between the two balances. In the Willyoung Apparatus

this is instantly accomplished by turning the spindle of a mercury "commutator" through 180°.

(2) Terminals of X and S should fit into mercury cups; contact resistances thus become negligible.

(3) The whole apparatus must be compact so as to have a uniform temperature throughout.

Although recorded as a text book method the "C. F. method" was practically unknown in this country until attention was directed to it by Mr. Willyoung in 1890 (see *Proc. Franklin Institute, Vol. II, pages 31 and 192*).* Since that time the method has been in constant use under Mr. Willyoung's hands and many improvements have been made. The pattern listed below is a "survival of the fittest" and embodies the very latest improvements which practically are of the apparatus here suggested.

W4240. **Willyoung Improved Carey-Foster Bridge and Commutator** $75 00

The bridge wire value may be varied by the use of shunts as originally suggested by Mr. Willyoung in 1892 (See *Proc. Electrical Section, Franklin Institute*, Vol. II., p. 31 and 192), thus requiring but one wire for all values of coils being worked upon and but one calibration instead of several as in the forms using changeable bridge wires. One commutation suffices to exchange both coils, while the battery may be independently commutated by use of a subsidiary head attached to the main commutator spindle. The copper work is massive and protected from amalgamation with the mercury by a surface of unbuffed nickel plate. The base is of hard rubber, thus furnishing a rigid substructure as well as the highest insulation. The bridge wire, which is of platinoid, is shielded from radiation of the body and convection currents by a strip of hard rubber, around which the contact key is curved, so that thermal E. M. F's are kept down. The apparatus is so designed that all loose mercury drains toward the center and into a pan beneath the base.

Fig. 32. W4240.

With this apparatus is supplied one shunt for the bridge wire and three auxiliary coils of 1, 10 and 100 ohms, without extra charge.

(*Cut is old and does not show recent improvements. Detailed blue print sent on application.*)

*These articles contain a most exhaustive treatment of the whole **Carey-Foster Method,** and should be read by those desiring a full acquaintance with the subject.

RESISTANCES AND COIL BRIDGES.

Units.—In the system of units now adopted for all scientific work the world over, and known as the C. G. S. system, the **Unit of Resistance** is defined as the resistance which is possessed by a conductor when unit difference of E. M. F. at its terminals produces unit current. This value is many times smaller than the values of resistance actually met with in practice, hence the **ohm** has been chosen as the practical unit of resistance. As defined in the C. G. S. system

One Ohm—10^9 C. G. S. units

For a long time past, various Electrical Congresses have been trying to frame a practical definition of the Ohm by means of which a resistance *equal to the ohm* could be made at any time by any skilled person, merely by "laying off" something to certain given dimensions. Mercury was early chosen as the substance and it was found that a column of about 100 cms length and 1 sq. mm. in cross section represented, roughly, one ohm. Successive Congresses now occupied themselves with the effort to fix the *exact* length at which the mercury column became an **Ohm.** We have, therefore, the following

	Length of Mercury Column in cms @ 0°c.	Date value was fixed.	Authority fixing this value.
The Siemens Unit	100 (1 sq. mm . C. S.)	1860	Dr. Werner Siemens
The B. A. Ohm	105	1864-5.	Brit. Ass. El. Stand. Com.
The Legal Ohm	106 (1 sq. mm. C. S)	1884	International Elec. Congress
The True Ohm	106.3 (" ")	1890	British Association
The International Ohm	106.3 ($\begin{smallmatrix}\text{must weigh}\\14.4521\text{ grams}\end{smallmatrix}$)	1893	International Elec. Congress

and we may find standards in use adjusted in any one of the above units.* The differences in ζ are obvious from inspection of the various *lengths* and must be taken account of in all but the roughest kind of practical work.*

From the above we may write

1 I. Ohm =1.01358 B. A. Ohm
1 B. A. Ohm= .9866 I Ohms

In a well equipped scientific laboratory and given the time of a well trained scientific worker, the mercury ohm can unquestionably be accurately produced. In the majority of cases however, it has been found safer to make copies of the mercury standard in the form of specially arranged coils of wire, to make these copies with the greatest possible care and refinement, and then to keep them as secondary standards for the production of others. Both the British Association and the Reichsanstalt (at Berlin) are now certifying in this way wire standards submitted them by manufacturers, and the time will doubtless quickly come, when in this country, we shall have a properly organized and legalized central bureau to perform the same service.

---

*We occasionally have complaint that **Sets** or **Standard Coils** are " off " by a ζ or so when received from us; we generally find that the purchaser is comparing with an old B. A. resistance differing from the **I. Ohm** by *over* **1.3ζ**!

Alloys.—In the manufacture of **Standard Resistances** the material used must have the following qualities:

(1) High Specific Resistance.
(2) Small change of resistance with varying temperature.
(3) Invariability of resistance with time.
(4) Not easily affected by chemicals or moisture.
(5) Small thermal E. M. F.

Pure metals answer only to (3) although (4) appears to hold for aluminum.

FREQUENTLY USED ALLOYS.

Composition.			Specific Resistance (Microhms per cubic cm.)	Temperature Co-efficient per Degree C
German Silver	(15%)	60 Cu + 25 Zn + 15 Ni	30	0.00036
" "	(30%)	60 Cu + 10 Zn + 30 Ni	38	0.00022
Platinum Silver		90 Ag. + 10 Pt	32	0.00027
Platinoid		30% G. S. with 2% tungsten added.		
Manganin		84 Cu + 12 Mn + 4 Ni	48	0.00003
Konstantan		60 Cu + 40 Ni	50	0.00004

Of the above German Silver has been most extensively used, but is now generally condemned because (3), invariability with time, is very badly met, particularly in moist climates or if subjected to much variation of temperature. This seems to be caused by mechanical disintegration of the wire due to its zinc component crystallizing out. *Distrust, therefore,* all alloys containing zinc. (*See Phil. Mag.,* Feb., '98.)

Manganin and Konstantan both seem ideal so far as Sp. Res. and Temp. Co-ef. go. Both are malleable and ductile. Konstantan has, however, an *excessive* thermal E. M. F. with copper, and is thus *absolutely prohibited* for high class standards. Manganin, in the other hand, is extremely oxidizable and must be treated with great care while being adjusted and mounted. It has seemed quite successful in Germany, although announced an absolute failure by English authorities.

Platinoid has been and is very largely used. Electrically it is nearly identical with 30% G. S., although chemically quite different; it contains platinum, silver, and 1 to 2% of tungsten. Observations in our own laboratory as well as reports which have come to us show that a properly prepared platinoid coil may with reasonable care in use and keeping be depended upon to vary not over a few 100ths % in, say, four or five years.

In the preparation of all of our resistances, we employ a special mode of treatment and ageing process the result of long experience with this kind of work. The aim is, first, to subject the coil to all the variations of temperature and condition which it will meet with in after life and, second, to permanently seal it against the action of moisture, gases, etc. We also accurately determine its temperature co-efficient, and in cases where several coils are put up in one containing box and must work together, we pick the wire used so that *each* coil may have the *same temperature co-efficient,* thus insuring the *relative* accuracy of the coils among themselves at *every* temperature.

Unless otherwise specified all the various Resistance Standards, Boxes, Wheatstone's Bridges, etc., which follow are wound with **Platinoid** wire. In adjusting, the coils are checked and rechecked in every possible combination and compared direct with first copies of original B. A. Standards and Reichsanstalt Standards for which we have the Cavendish and Reichsanstalt certificates respectively.

Z. C. Alloy.—We invite attention to our "Z. C" Wire—this wire has a practically negligible temperature co-efficient, and a specific resistance about 1.5 times that of G. S. We have been making tests upon it for a long period now and can recommend it very highly as a very reliable material for resistance work when prepared as we know how to prepare it.

GUARANTEE AND CERTIFICATE.

With each of our various Resistance Coils, Boxes, Wheatstone's Bridges, etc., we furnish a certificate signed by the firm and our chief adjuster and giving the exact resistance at a given temperature, the accuracy of adjustment, and the temperature co-efficient. This certificate also contains our guarantee of perfect workmanship, and binding ourselves to make good any errors of or changes in value without charge except that of transit to and from our laboratory, if reported within one year from purchase, and not due to neglect or violation of our directions as to the proper use of the instrument.

Fig. 33. W4246-48.

SINGLE STANDARDS OF FAIR ACCURACY.*

W4246. **Standard Resistance**............ ....$5 00
Any desired value between 0.5 and 100 ohms. Mounted in polished hard rubber block and has both strip (for meter bridge) and binding post terminals. Accuracy 1-10%.
W4247. **Standard Resistance**.............................. $6 00
Same as W4246, but any value between 100 and 1000 ohms.
W4248. **Standard Resistance**.............................. 8 00
Same as W4246, but any value between 1000 and 4000 ohms.
W4249. **Standard Ohm**.... ....Grade A 27 50
In four sub-divisions of 0.1, 0.2, 0.3, and 0.4 ohms. In rubber topped mahogany box. Accuracy 1-10%.
***In Stock.** We try to carry in stock (W4246-W4248) coils of 0.5, 1, 10, 100 and 1000 ohms. All others must be wound to order.

SINGLE STANDARDS OF HIGH ACCURACY.

(B. A. FORM.)

This is the oldest recognized form of Single Standard, having been adopted by the British Association in 1864. It is still largely used. For work of the very highest accuracy, however, the Reichsanstalt style of mounting is probably superior.

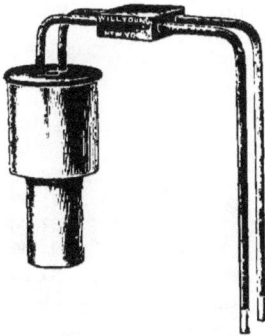

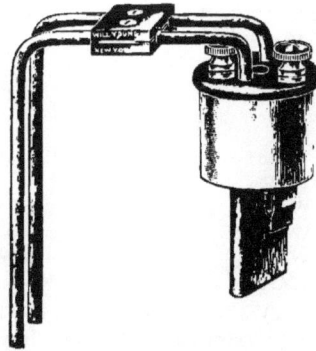

Fig. 34. B. A. Ohm.   Fig. 35. B. A. Ohm (Chrystal's Form).

The B. A. form is rather slow to take up changes in temperature; it must be left in its oil or water bath same time, therefore, to be certain that it itself has the bath temperature. If supplied with Prof. Chrystal's thermo-couple, one leg of which is embedded in the coil, while the other leg dips into the bath, this equality of temperature is at once observable by joining a galvanometer to the circuit of the couple.

Accurate to 1-100%.

W4253.	**Standard International Ohm**............Grade A $25 00
W4254.	" " 10 Ohms......... " 27 00
W4255.	" " 100 Ohms....... " 28 00
W4256.	" " 1000 Ohms ...... " 30 00
W4257.	" " 10000 Ohms ..... " 32 00

Each of the coils W4253-57 is supplied in a polished mahogany case with lock and key.

For any coil W4253-57 with Chrystal thermo-couple, extra, $5.00.

Cavendish Certificates. On any of the coils W4253-57, either ordinary or Chrystal's form, we shall be glad to quote Duty Free Prices to Educational Institutions for English manufacture; in this case a Cavendish laboratory certificate can be furnished for a small additional charge. Three to six months must be allowed, however, for delivery.

SINGLE STANDARDS OF HIGH ACCURACY.
(REICHSANSTALT FORM)

The "Reichsanstalt" Series of Single Standards has been developing during the past six or seven years, under the direction of the workers in the Imperial Laboratory, Berlin. Except in a few of the Standards for very heavy currents the alloy is "manganin" wound in shellac upon metal cylinders or other forms of supporting structure and baked hard. No final "sealing in" with paraffin is employed, and, in use, an oil bath is employed, the mountings being so arranged that the oil comes in direct contact with the wire.

These Standards are to be recommended as simple, practical and substantial. They are unquestionably superior to the "B. A." form.

TYPE A.
Accurate to 1-100%

W4265.	**Reichsanstalt 1-10 Ohm Standard**			....	$20 00
W4266.	"	1	"	"	.... 20 00
W4267.	"	10	"	"	.... 21 00
W4268.	"	100	"	"	.... 22 00
W4269.	"	1000	"	"	.... 24 00
W4270.	"	100000	"	"	.... 35 00

W4265-69 are mounted in Nickel-Plated Brass cases 8 cm in diameter with heavy copper leads to drop into mercury cups; no binding post terminals. They will carry currents up to 1 ampere without injury when used with the oil bath.

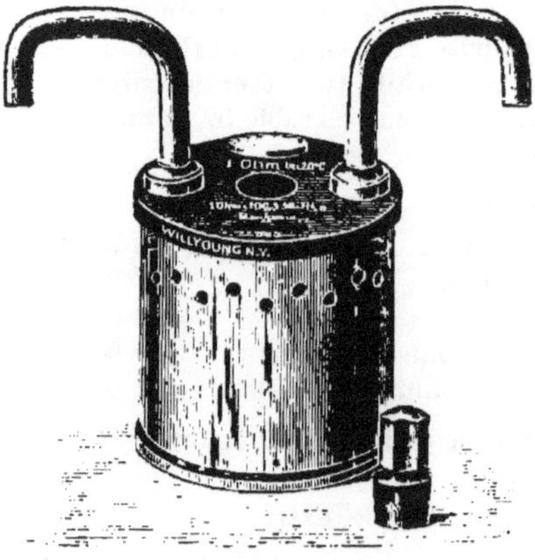

Fig. 36. REICHSANSTALT STANDARD. TYPE A.

Fig. 37. OIL BATH. W4270.

TYPE B.

W4271. **Reichsanstalt 1-10 Ohm Standard**..............$25 00

W4272.   "   **1**   " 30 00

Type B differs from Type A in that currents up to 4 amperes may be carried: the copper leads and resistance wire proper are heavier and the mounting correspondingly larger.

W4273. **Oil Bath**..$20 00

For any resistance W4265-72 inclusive. Of hard soldered and nickel-plated copper. Has stirrer and faucet for drawing off the oil.

W4274. **Oil Bath**................................... Grade B 25 00

For W4265 72 inclusive. Four resistances may be placed in series or parallel at one time; with stirrer and faucet for drawing off the oil.

W4275. **Oil Bath**...................................Grade B $30 00

Same as W4274, but for five resistances instead of four.

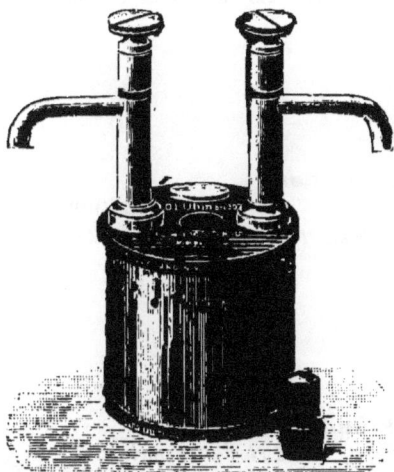

Fig. 38. REICHSANSTALT STANDARD. TYPE C.   Fig. 39. REICHSANSTALT STANDARD. TYPE D.

LOW RESISTANCE STANDARDS FOR CURRENT MEASUREMENTS.
(REICHSANSTALT FORM)

These resistances are especially intended for accurate current measurements by the " Fall of Potential" method, which consists in solving the equation $c = \frac{E}{R}$ from a knowledge of the E. M. F. produced at the terminals of a *known* resistance, and the flow of a steady current.

TYPE C.

W4281. **One-tenth Ohm Standard** .... Grade B $25 00
   Capable of carrying 10 amperes.   Accuracy 1·25%.

W4282. **One One-hundredth Ohm Standard** ...   "   30 00
   Capable of carrying 100 amperes.   Accuracy 1·10%.

W4283. **One One-thousandth Ohm Standard** ...   "   75 00
   Capable of carrying 250 amperes; a pair of massive lugs is provided
into which cables carrying the current to be measured may be soldered.
Accuracy 1·5%.

   W4281-83 have very massive copper leads to dip into mercury cups;
a pair of special "potential" terminals are separately carried *direct to the
terminals* of the resistance proper.

TYPE D.

W4284. **Reichsanstalt 0.0001 Ohm Standard**
                                              Price on Application
   Has both Mercury Cup and Binding Post Terminals; carries 200
amperes.

TYPE E.

W4285. **Reichsanstalt 0.001 Ohm Standard**
                                              Price on Application
   Carries 400-500 amperes.   A small turbine permits oil to be continuously stirred by small motor.

W4286. **Reichsanstalt 0.001 Ohm Standard**
                                              Price on Application
   Carries 1000 amperes.   Design and Construction similar to W4285.

W4287. **Reichsanstalt 0.0001 Ohm Standard**
                                              Price on Application
   Similar to W4285-86 but larger.   Carries current up to 2000 amperes.

   Reichsanstalt Certificates.—Any of the here listed Reichsanstalt Standards can be supplied (made by Wolff of Berlin) with Reichsanstalt Certificates of accuracy, etc., at a very trifling advance (to cover the
Reichsanstalt fee).   *Duty free prices to Colleges and Educational Institutions will be given on application.*

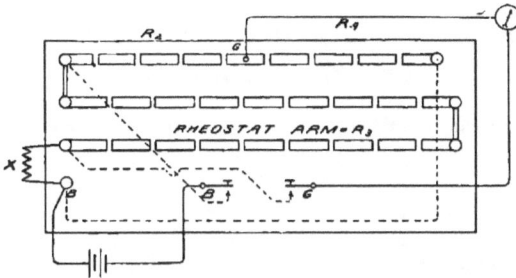

Fig. 40. SKELETON OF W4296.

The Coil Bridge differs from the Slide Bridge in having the various resistances (R_2, R_3, and R_4 of fig. 27) made up of a number of individual coils so arranged that the total of each Bridge Arm may be changed at will by means of plugs or switches. Fig. 27 shows in skeleton our W4296. The upper row constitutes the Bridge or "ratio" Arms R_4 and R_2 with coils of 1, 10, 100 and 1,000 ohms on each side of the centre, while the two lower rows taken together make up the "rheostat" arm R_3. In use one coil of each Bridge Arm is left unplugged ; if we have 10 out in the left arm and 100 out in the right and the balancing resistance in the rheostat arm is R, then we may write

$$\frac{X}{R} = \frac{10}{100} \text{ or } X = \frac{1}{10} R$$

If a change in R will not give a balance, the ratio coils R_2 and R_4 must be suitably altered.

Fig. 41. W4296.

RESISTANCE BOXES AND WHEATSTONE BRIDGES OF FAIR ACCURACY.

Wound with Platinoid Wire and adjusted to an accuracy of 1-5%. In cherry box with segments mounted on hard rubber. Substantially constructed.

W4290, **Resistance Box**............................Grade B $32 50
1111,5 ohms, divided into fourteen coils of 0,5, 1, 1. 2, 3, 4, 10, 20, 30, 40, 100, 200, 300 and 400 ohms.

W4291. **Resistance Box**............................Grade B $45 00
Same as W4290, but with addition of 1000, 2000, 3000 and 4000 ohm coils, making a total of 11111.5 ohms.

W4295. **Resistance Box and Wheatstone Bridge**
Grade B $50 00
Fourteen coils of platinoid wire—0.5, 1, 1. 2, 3, 4, 10, 20, 30, 40, 100, 200, 300 and 400 ohms; total resistance 1111.5 ohms. Also Bridge Arms 1, 10, 100 and 1000 ohms on each side. and keys for battery and galvanometer.
Bridge coils accurate to 1-10%.

W4296. **Resistance Box and Wheatstone Bridge**
Grade B $65 00
Same as W4925, with addition of four rheostat coils—1000, 2000, 3000 and 4000 ohms, making eighteen coils with total resistance of 11111.5 ohms.

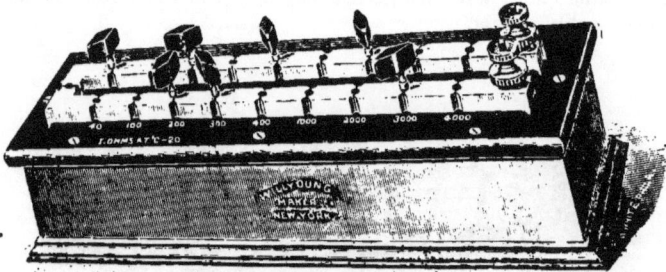

Fig. 42. W4305. Grade A.

RESISTANCE BOXES AND WHEATSTONE BRIDGES
OF HIGH ACCURACY.

The following Resistance Sets are strictly high grade, and no time or expense is spared to make them the very best of their types. All wire is carefully selected, and the hard rubber used is *specially made for us* and warranted free from metallic or other impurities.

W4300. **Standard Resistance Box and Wheat-stone Bridge,** (Anthony Form).......... Grade A $400 00 " B 350 00

The coils are of platinoid wire, divided as follows: ten one-tenth ohm coils, ten units, ten tens ten hundreds, and ten thousands—11111 ohms total resistance. They are arranged in rows and may be joined in series or multiple or in any combination of series and multiple desired. Bridge ratios consist of 1, 10, 100, 1000. and 10000 ohms in each arm; these arms may be reversed by means of a simple plug device attached to the bridge. A copper coil, embedded among the other coils of the set, is so arranged that by a plug it may be substituted for the usual unknown resistance, and its resistance measured in terms of the coils of the set. It changes one ohm in apparent resistance as so measured for each change of 1°C in temperature, thus enabling the real temperature of the coils to be found far more accurately and conveniently than by use of the mercury thermometer. Brass work is very massive, as are also the copper connections beneath the set. Segments are so cut as to make thorough cleaning easy. The coils in rheostat are guaranteed accurate to 1·50% and in bridge to 1·100%.

Fig. 43. W4300.

W4301. Standard Resistance Box and Wheatstone Bridge, (Improved P. O. Pattern... Grade A $225 00 " B 210 00

This box is nearly three times the size of the usual English P. O. Pattern, thus permitting the use of large wire (reducing heat effects) and brass segments of greater mass and more insulating space. Coils are of platinoid, as follows:—0.5, 1, 2, 3, 4, 10, 20, 30, 40, 100, 200, 300, 400, 1000, 2000, 3000, 4000, and 10000 ohms; total resistance 21110.5 ohms. Bridge ratios are 1, 10, and 100 ohms on one side, and 10, 100, and 1000 ohms on the other, with a special plug reversing device, which renders these six coils equivalent to the usual EIGHT. Battery and galvanometer keys are mounted upon the set and two traveling plugs for isolating particular coils are provided; Rheostat coils are guaranteed to 1-25%, and bridge arms to 1-50%.

W4302. Standard Resistance Box and Wheatstone Bridge......................... Grade A $130 00 " B 110 00

Eighteen coils of platinoid wire:—0.5, 1, 2, 3, 4, 10, 20, 30, 40, 100, 200, 300, 400, 1000, 2000, 3000, 4000, and 10000 ohms—total resistance of 21110.5 ohms. Bridge arms 1, 10, 100, and 10, 100, 1000 ohms; with reversing devise to secure same result as from eight coils. Galvanometer and battery keys, also two traveling plug binding posts for isolating particular coils. Rheostat coils accurate to within 1-25 of 1%; bridge coils accurate to within 1-50 of 1%. *This set is exactly the same as W4301, save that it is about ⅓ smaller.*

W4303. Standard Resistance Box............... Grade A $80 00 " B 60 00

111.5 ohms total resistance, divided as follows:—0.5, 1, 1, 2, 3, 4, 10, 20, 30, and 40 ohms. Of large sized wire and massive blocks. In rubber topped mahogany case. Accuracy 1-25%.

W4304. Standard Resistance Box............... Grade A $100 00 " B 75 00

Same as W4303, but with additional coils of 100, 200, 300, and 400 ohms. Total resistance 1111.5 ohms.

W4305. Standard Resistance Box............... Grade A $125 00 " B 90 00

Same as W4304, but with additional coils of 1000, 2000, 3000, and 4000 ohms. Total resistance 11111.5 ohms.

W4306. Standard Wheatstone Bridge (without Rheostat)..................................... Grade A $100 00 " B 75 00

Same style as W4303-W4305. Arms of 1, 10, 100, and 1000 ohms on each side and with reversing arrangement as in B5230-B5231, etc. Accuracy 1-50%.

Traveling Plugs for W4301-W4305 ....per pair $1 00

W4310. Centigrade Thermometer........................$5 00

Mounted with bulb inside of box for Nos. W4301 to W4305 inclusive.

Fig. 44. W4301.

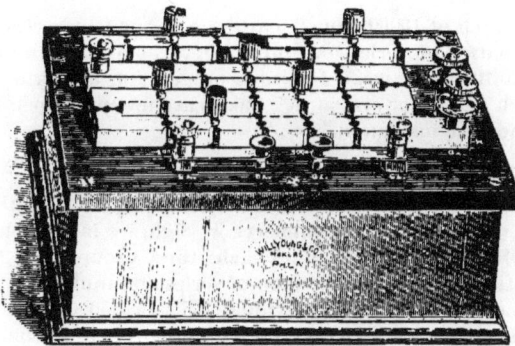

Fig. 45. W4302. GRADE B.

STANDARD HIGH RESISTANCES.

W4315. **50000 Ohms**...............................Grade B $25 50

In brass case with rubber top ; two terminals at top of hard rubber pillars.

W4316. **100000 Ohms**...........................Grade B $32 50

Same style as W4315.

W4317. **250000 Ohms**...........................Grade B $75 00

Five groups of coils of 50000 ohms each. Same style of mounting as W4316.

W4318. **500000 Ohms**...............................Grade B $125 00

Same as W4317, but 100000 ohms in each group.

W4320. **Megohm Box**...........................Grade B $200 00

Same as W4317, but 200000 ohms in each group.

W4325. **Resistance Box**, 100000 ohms...............Grade B $35 00

Mounted in brass case with rubber top. Connections should be made to terminals marked 3 and 4 When the flexible cord is on plug 1 the box is short circuited; but when on plug 2 the resistance of 100000 ohms is in series. Specially suited to rapid cable testing.

W4330. **Resistance Box**, 100000 ohms, (square pattern). Grade A $67 50
" B 50 00

In four units of 10000, 20000, 30000, and 40000 ohms. An "infinity" plug separates each coil from the one next to it. Segments are elevated upon the hard rubber top by special washers to increase insulation. *Binding posts are so arranged as not to be in the way when plugs are used, a common defect of most square pattern sets.*

W4332. **Resistance Box**, 100000 ohms, (long pattern). Grade A $125 00
" B 100 00

Ten coils each of 10000 ohms; may be arranged in series or multiple and in combination of series and multiple, to obtain resistances varying from 1000 to 100000 ohms. Is also arranged for use as a Wheatstone Bridge of high resistance arms. All blocks are mounted upon hard rubber washers to increase insulation.

W4333. **Megohm Resistance Box**.............. Grade A $325 00
" B 275 00

The design of this box is similar to that of W4332 save that the bars are mounted upon hard rubber supports, 4" high, for insulation. The coils are in ten (10) groups of 100000 ohms each; these groups may be joined in series or multiple, and in combination of series and multiple. Each group is actually made up of four (4) individual coils of 25000 ohms each. The rubber supports are drilled out and *do not touch* the rods connected with coils except at top. The insulating surface is thus very long. *As designed by Mr. Willyoung and made for Prof. Wm. A. Anthony, of the Cooper Union.*

W4315 to W4333 inclusive are accurate to 1-25%.

Fig. 46. W4325.

Fig. 47. W4330.

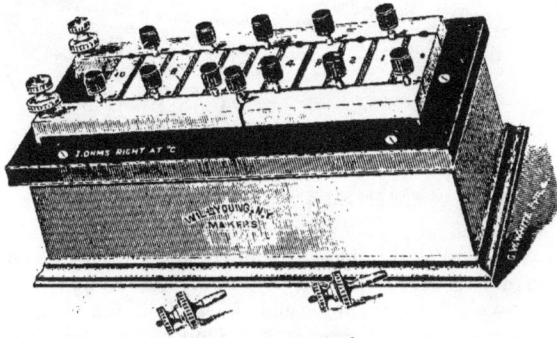

Fig. 48 W4332.

Fig. 49. W4333.

"AONE" PORTABLE TESTING SET •

For all Kinds of Electrical Testing; Invaluable to the Scientist or the Engineer, in the Laboratory or the Shop.

IMPORTANT TO KNOW.

Although one of the most ancient of test instruments, the **Wheatstone's Bridge** still remains one of the most valuable and useful to the electrician and engineer having electrical measurements to make. This assumes, however, that the bridge is well designed and conveniently arranged, and that, furthermore, it includes the necessary battery, galvanometer, range of resistance coils, and range of bridge ratios.

The "Aone" is the most recently designed of all Portable Sets upon the market, and is the personal work of our Mr. Willyoung, whose experience has been, perhaps, larger in the production and use of Sets of this class than that of any other worker in this country. It is believed that the most carping critic will be unable to find anything lacking in this set for the accurate and convenient carrying out of all the various tests for which its usefulness is claimed.

DESIGN AND CONSTRUCTION.

It is the smallest "Testing Set" ever made—measures 8¾"x 5"x 4" deep.

It is the lightest "Testing Set" ever made—weighs less than 7 pounds.

*It has a wide range of measurement—0.001 ohm to over 11 megohms.

---

* These are limits as computed from the values of the minimum and maximum rheostat resistance and the maximum bridge ratios. Owing to the impossibility of constructing a portable galvanometer of sufficient sensitiveness, the actual accurate working limits are from about 0.01 ohm to about 300,000 or 400,000 ohms, unless a greater battery power be employed.

It is complete in itself, with its own battery and galvanometer.

It is *absolutely unaffected* by mechanical or electrical disturbances, or by neighboring electrical currents.

It does not require leveling; may be set down and used anywhere.

It has one of the most sensitive galvanometers ever made, with an external adjustment by means of which the needle may be always kept on zero, despite any slight changes in the controlling springs.

WHAT THE "AONE" PORTABLE TESTING SET WILL DO.

(1) In general, measure any unknown resistance.

(2) May be used to compare E. M. F.'s, *i.e.*, measure an unknown E. M. F. in terms of a known E. M. F.

(3) May be used to check an ammeter.

(4) May be used to check a voltmeter.

(5) May be used to measure a battery (internal) resistance.

(6) May be used to measure insulation resistance by the "direct deflection method."

(7) Will accomplish the "Murray Loop Test."

(8) Will accomplish the "Varley Loop Test."

(9) May be used as an ordinary resistance box—bridge and rheostat in series.

(10) May be used as a resistance box—rheostat only.

(11) Galvanometer may be used separately.

(12) Galvanometer and battery (in series) may be used separately.

(13) May be used with an outside E. M. F. in place of, or added to, the regular battery.

(14) May be used to test out "short circuits" of all kinds.

DESCRIPTION.

The "Aone" Portable Testing set is, essentially, a regular Wheatstone Bridge, including battery and galvanometer, with all the parts designed to stand a maximum of wear and tear while occupying a minimum of space. Convenience of operation with simplicity of parts has also been sought.

The Rheostat Arm has 16 coils of resistance, 1, 2, 3, 4, 10, 20, 30, 40, 100, 200, 300, 400, 1,000, 2,000, 3,000 and 4,000 ohms—11,110 ohms total. They make up the two lower rows (fig. 51), being joined by a heavy upper rod beneath.

Bridge Arms (upper row, fig. 51) are 1, 10, 100 on one side and 10, 100 and 1,000 on the other.

Reversing Arms or "Commutator" is the name applied to the four blocks A, B, X and R taken together. By placing plugs in one diagonal or the other (fig. 53 or fig. 54) the two bridge arms are transposed, with reference to one another, so that if in one case $X = \dfrac{A}{B} R$ in the other it will be $X = \dfrac{B}{A} R$. (This may be nicely followed out in fig. 52.)

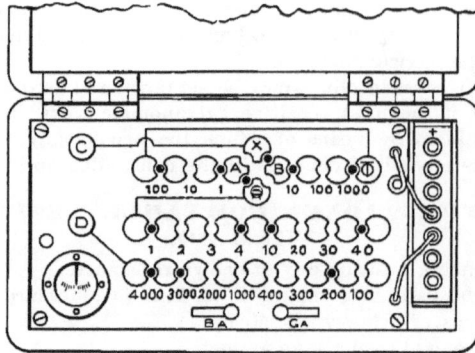

Fig. 51. PLAN OF "AONE" TEST SET.

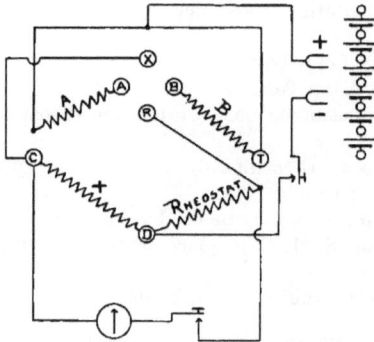

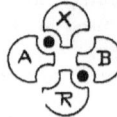

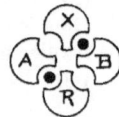

Fig. 52.   Fig. 53.   Fig. 54.

The Galvanometer is a very sensitive instrument of the D'Arsonval type. The coil is pivoted in sapphire jewels and is "dead-beat" without the use of a "short circuit" key. The scale is long and the deflection strictly proportional to current, so that the instrument may be used for insulation measurements by the "Direct Deflection" method if desired. The instrument is thoroughly portable and not easily gotten out of order; *it is not disturbed by neighboring magnetic fields or electrical circuits, and may be used in proximity to the heaviest electrical machinery.*

A small external milled head permits the galvanometer zero to be restored should its control springs ever change their value.

The Battery.—Six silver chloride of cells are mounted in a separately detachable box. The flexibles permit one or more to be placed in series. These flexibles allow of the insertion of outside battery if desired.

Finish.—The "Aone" Set is contained in a finely polished mahogany case with nickel plated handle and a *thoroughly good lock and key*. The handle is on the side, which is not only the most convenient position for carrying, but makes it impossible for the whole bottom to drop away from the lid and thus, possibly, do serious injury. The brass work is finely lacquered and the hard rubber well finished. Material and workmanship is of the best in every way.

Accuracy of the rheostat coils is $\frac{1}{3}\%$. and of the bridge coils $\frac{1}{10}\%$.

Full directions as to the use of the "Aone" Set in making the various measurements is furnished with the set.

PRICE LIST.

W4340. **"Aone Portable" Testing Set**.................. $75 00

W4341. **Leather Case** for same, with handles and sling strap.   5 00

The following are among the users of the "Aone" Testing Set :

*THE U. S. NAVY,   CRAMP'S SONS & CO.,

U. S. WAR DEPARTMENT,   NEWPORT NEWS SHIPBUILDING & DRY

JAPANESE NAVY,   DOCK CO.

Interested parties should write for our special circular, "The 'Aone' Portable Testing Set," in which the description, etc., is made more elaborate than space will here permit of.

GREELEY TEST SETS.

Y1 AND Y2.

This well-known set has been upon the market for a number of years, and is very widely used by those engineers who are associated with or have grown out of telegraph interests. While perhaps not theoretically as perfect as some of the more recently designed sets, yet the fact that this set is so well known and has proved so trustworthy has led us to continue its sale. While a number of minor improvements have been made, the set is still, nevertheless, the orthodox set.

Fig. 55.   Y1 (NO BATTERY).

* During the recent war with Spain we equipped a large number of the newly purchased boats with these sets.

Y1. Bridge and Testing Set............................ $90 00

This instrument has rows of unit, ten and hundred ohm coils in rheostat, and ten and hundred ohm coils in each arm of the bridge. Range of measurement is from $\frac{1}{10}$ of an ohm to 11,100 ohms. Coils of platinoid and accuracy of adjustment $\frac{1}{5}$ for rheostat and $\frac{1}{10}$ for bridge coils. Has battery of five chloride of silver cells. In mahogany case, with lock and key and handle.

Y1a. Bridge and Testing Set........................... $80 00

Same as Y1, but without battery.

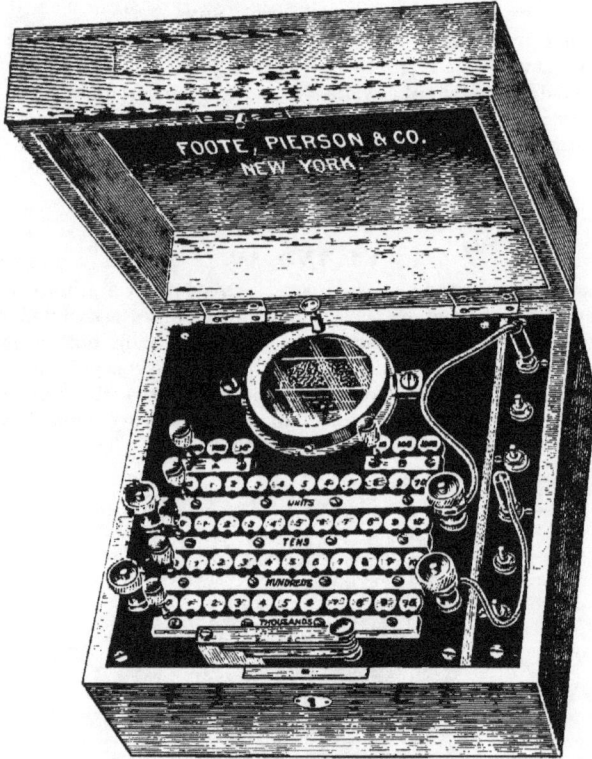

Fig. 56. Y2.

Y2. Bridge and Testing Set............................ 105 00

Similar to Y1, but with three coils of 10, 100 and 1000 in each bridge arm. Rheostat, also, has four rows of ones, tens, hundreds and thousands respectively. Has battery of five chloride of silver cells.

Y2a. Bridge and Testing Set... .... $95 00

Same as Y2, but without battery.

RESISTANCE BOXES AND WHEATSTONE BRIDGES.
"NINK" (NON-INDUCTIVE—NON-CAPACITY.)

For all measurements where standard Resistances are required, but especially useful for self-induction work, and whenever alternating currents are used.

As ordinarily made resistance coils are double wound; this eliminates self-induction, but leaves large capacity so that such coils cannot be used in making self-induction determinations by any of the ordinary methods involving resistances.

In the "Nink" Boxes every coil is composed of two wires in multiple, each of which wires has double the resistance of the finished coil. Each wire is wound into a flat and very thin coil. The two coils are then placed one flat upon the other and with windings running in the opposite direction. The outer and inner pair of terminals are then joined so as to place the two coils in parallel. It is now obvious that the coils are still non-inductive since the current is evenly split and passes in opposite directions while the potential from periphery to the center is the same in one half as in the other. As a condenser, therefore, this completed coil has a capacity equal to that of a condenser whose two surfaces are the inner and outer peripheries respectively, the distance between the condenser surfaces being the distance between these peripheries and the potential taken as half the difference of potential existing between the outer and the inner terminal. Both experiment and theory have shown that this capacity even for coils of high resistance is exceedingly small and may be taken, for all purposes of measurement, as nil.

The expense of winding and adjusting the "Nink" coils is considerably greater than that of the usual type; we are prepared to furnish the *Resistance Boxes and Wheatstone Bridges described in preceding pages, with "Nink" coils, at prices that will be quoted upon application.*

RESISTANCE BOXES, BRIDGES, ETC., WITH ZERO TEMPERATURE COEFFICIENT.

Any of the preceding Resistance Boxes, Bridges, etc., can be supplied wound with our "Z. C." Alloy, having negligible change of resistance with temperature ; owing to the costliness of this wire as compared with other orthodox alloys, the prices of apparatus employing it are necessarily increased, and are, therefore, not given here, but will be furnished on application. For the same reason no apparatus thus wound is kept regularly in stock, and from two to three weeks from date of order must always be allowed.

We have tested this alloy thoroughly and during a period sufficiently long to be sure that, processed as we process it, it is unchanging and in every way thoroughly reliable and free from all objectionable features.

KEYS.

W4350. **Single Plug Key**, hard rubber base .... Grade A $5 00
" B 4 50

W4351 **Double** " " " " A 7 00
" B 6 25

W4352. **Three-way** " " " .... " A 6 00
" B 5 50

For above keys with blocks mounted upon hard rubber pillars for high insulation and with spring-capped plugs (to prevent blocks being forced apart where plug is inserted), double the price. Fig. 58 illustrates W4351 mounted in this manner.

Fig. 57. W4351.

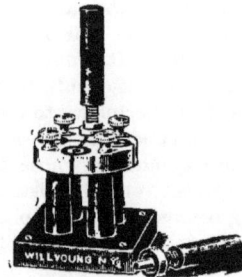

Fig. 58. W4351 (In Pillars.)

Fig. 59. W4356.

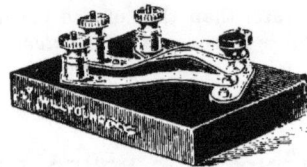

Fig. 60. W4357.

W4356. **Single Contact Key** on hard rubber base..... Grade A $6 00
" B 5 00

W4357. **Double** " " " " " A 10 00
" B 8 00

W4358. **Short Circuit** " " " ..... " A 6 50
" B 5 75

W4359. " and Single Contact Key with improved eccentric "hold down".............. " A 10 00
" B 8 00

The hard rubber eccentric is held by a friction spring washer. Varying the handle permits permanent closure of the circuit or the use of the key as a plain single contact key; or, throwing lever clear back instantly throws the upper—galvanometer "short-circuit key"—contact in function.

W4360. **Short Circuit Key, etc.**.................... Grade A $20 00
                          " B 16 00

Same as W4359 but on pillars, thus greatly increasing the insulation. *We Recommend* this style for permanent insulation testing outfits.

Fig. 61. W4359.

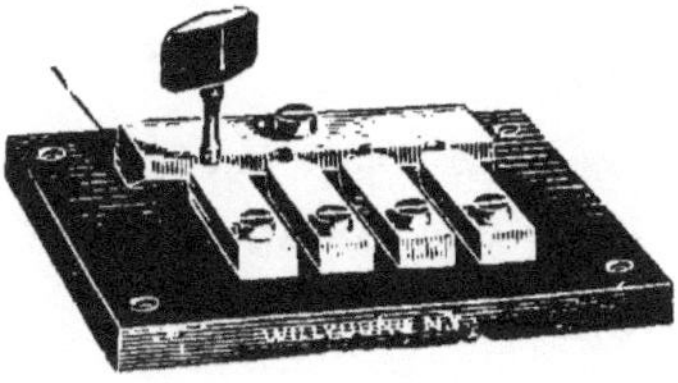

Fig. 62. W4362.

W4361. **Battery Switch,** 2 connections........................$ 7 50
W4362.   "   "   4   " 12 00
W4363.   "   "   6   " 17 50

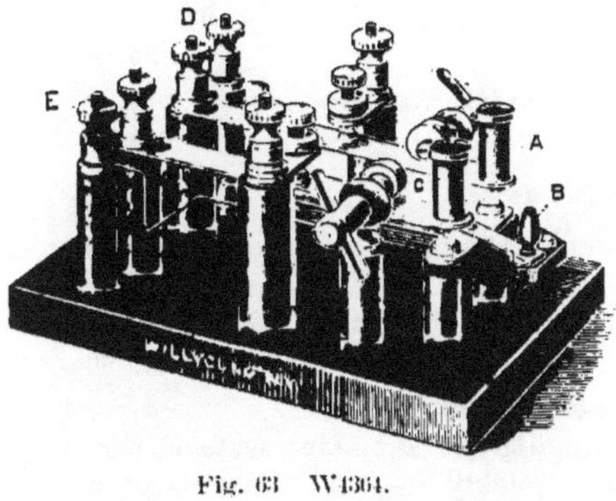

Fig. 64. W4366.

Fig. 63. W4364.

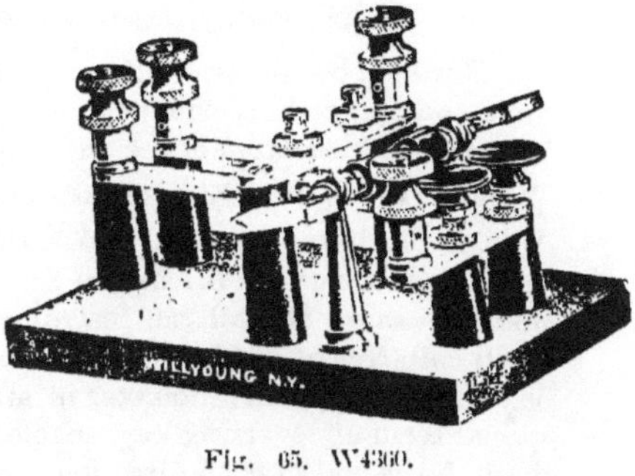

Fig. 65. W4360.

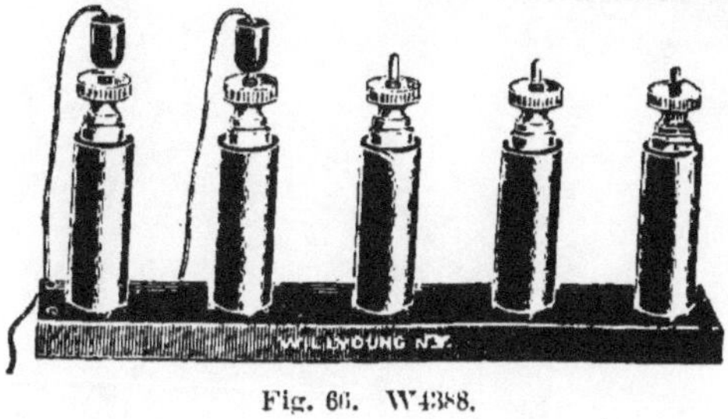

Fig. 66. W4388.

W4364. **Willyoung's Improved Key**..............Grade B $27 50

For insulation testing, etc.; combines short circuit key, circuit closing key, condenser discharge key and battery changing key.

W4366. **Lambert's Discharge Key**............... Grade A $25 00
"   B   20 00

W4368.   "   "   " .... " A 30 00
"   B   25 00

Same as W4368, but with eccentrics to hold key down.

W4370. **Reversing Key**.......................... Grade A $30 00
"   B   25 00

W4372.   "   " .... " A 35 00
"   B   30 00

Same as W4370, but cut through.

W4374. **Reversing Key**.......................... Grade A $40 00
"   B   35 00

Special pattern constructed to secure exceedingly high insulation.

W4376. **Lambert's Capacity Key**................. Grade A $30 00
"   B   25 00

(See *Kempe*, p. 350.)

W4378. **Double Connectors on Ebonite Pillars,** 2 posts, $9 00
W4380.   "   "   "   " · 3 " 15 00
W4382.   "   "   "   " 4 " 20 00
W4384.   "   "   "   " 6 " 26 00

W4386. **Set of Three Cable Posts with Connectors**.... 7 50
W4388. " **Five** "   "   "   " 12 50

W4389. **High Insulation Line Posts**......... each, 1 50
per half doz., 15 00

Hard rubber pillars with wood screw at bottom; they screw into test table and carry the permanent wires of the test table at their top. thus keeping the insulation of the entire outfit at the highest possible limit.

W4390. **Kelvin's Electrometer Key**...................... 22 50

W4391. **Damping Coil and Checking Key** 30 00

This is a very useful device for bringing galvanometer systems, particularly those of ballistic instruments, quickly to rest. Consists of a small coil (free of magnetic material) so mounted upon a stand that it may be presented to a galvanometer in any desired relation. A dry cell and a special form of reversing key enables two strengths of current to be put through the coil in each direction.

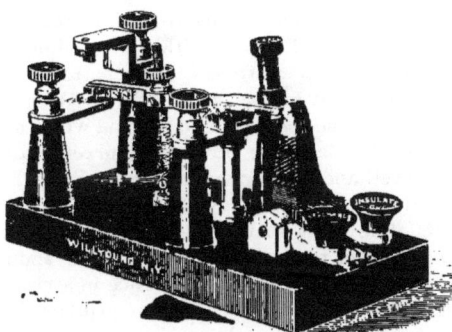

Fig. 67. W4400.

KEYS WITH RUBBING CONTACTS.

W4392. **Short Circuit Key**............. ....Grade B $17 50
On high insulation rubber pillars.

W4393. **Improved Reversing Key**...............Grade B $40 00

W4394. **Muirhead's Reversing Key**..............Grade B $50 00
(See *Kempe*, p. 286.)

W4396. **Rymer-Jones Discharge and Reversing Key**
Grade B $40 00

(See *Kempe*, p. 286.)

In the Rymer-Jones key the contacts are all rubbing contacts, the switches moving in a horizontal plane; whatever contact is made is, hence, *well* made and *permanently* made as well. The functions of the orthodox Reversing key and of the Discharge key are *both* performed by this key. A further feature of the key is that wrong connections *cannot* be made with it—battery short circuit or battery dead on galvanometer *e. g.* The design is new, the insulation very high, and the key as a whole is most heartily recommended.

W4398. **Saunder's Improved Capacity Key**.....Grade B $55 00
(See *Kempe*, p. 356.)

W4400. **Kempe's Discharge Key**.. .... Grade A $50 00
" B 35 00

With rubbing contacts. This key is a great improvement over previous forms. The contacts are always bright and clean, while the insulation is of the highest.

STANDARDS OF E. M. F.

Definition. Certain forms of primary batteries made with chemically pure materials according to certain specified methods have been proven to give a certain definite and fixed E. M. F. at a given temperature, and to maintain this same E. M. F within exceedingly small limits of change over a long interval of time. Such batteries are spoken of as **Standard Cells** and may properly be considered as Electrical Instruments. They must only be used in some method which allows the smallest possible current to flow, since the flow of a current causes more or less polarization of the cell and consequent change of E. M. F.

Caution. It is best to employ Standard Cells in connection with a *"zero method,"* like the potentiometer methods, later to be discussed, which allows *no* current to flow. In any case one should *never be used* except with at least 10000 ohms extra resistance in series with it.

The Daniell Cell. This well-known primary battery makes an excellent "standard cell" if properly prepared. It is, however, not portable.

For laboratory purposes a very effective and simple form may be quickly put together as below (fig. 68.)

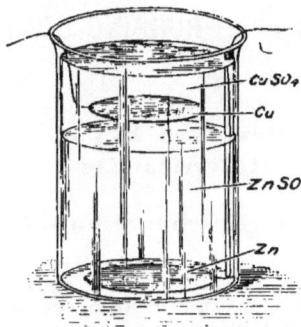

Fig. 68. DANIELL STANDARD CELL.

The zinc plate must be chemically pure and freshly amalgamated; a wire is soldered to it, covered with a glass tube, and passes up through the liquids. The jar is filled about half full with a solution (density 1.2 @ 15° C) of chemically pure $ZnSO_4$ in distilled water Upon this is poured a solution (density 1 2 @ 15° C) of chemically pure $Cu SO_4$ in distilled water. In the $CuSO_4$ is then hung a disc of copper with wire terminal. This copper may be ordinary copper of commerce, but must be freshly electroplated just before use.

The two solutions must lie sharply upon one another without diffusion. To accomplish this, place a disc of stiff note paper upon the $Zn SO_4$ before pouring on the $CuSO_4$; then pour the latter gently through a funnel, the tip of which is nearly in contact with the paper. As the $CuSO_4$ is poured, the paper will rise with it and the two solutions will lie one upon the other without mixing.

The E. M. F. of a Daniell Cell so prepared is 1.07 volt and the change of E. M. F. with temperature may be neglected. Many other forms of Daniell Cell, some much more elaborate, have been suggested. The requirements are the same, however, in all, viz., *chemically pure materials, freshly amalgamated zinc, and a sharp plane of separation between the two solutions.* This means that the cell must practically be freshly set up each time it is used.

U. S. CONGRESS CELL.

This cell (fig. 69) has been made the legal standard of E. M. F. for the U. S. by act of Congress and on recommendation of the National Academy of Sciences. It is a form of Clark cell having two communicating "legs' and employs material prepared and inserted according to a specification also legalized. The cell is not portable, but must be prepared by the user.

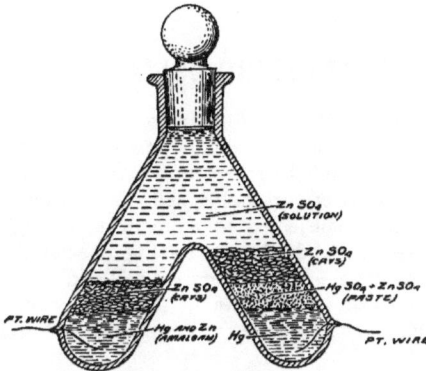

Fig. 69. W4390.

Fig. 70. W4395.

U. S. CONGRESS STANDARD CELL.

W4390. **Glass Cell Only**............................ .... $4 50
  With platinum leading in wires; ready for mounting up.
W4391. **Six Glass Cells** 22 50
W4392 **Material for Two Cells**........................... 4 50
  Consisting of
    (1) Bottle of mercury with dropper,
    (2) "   amalgam "   " and tube,
    (3) "   mercurous sulphate,
    (4) "   zinc sulphate and tunnel,
    (5) "   shellac and brush,
    (6) Glass rods, pusher, etc.,
    (7) Copy of specifications and directions for setting up.

W4393. **Glass Cell, Etc.**...................... .... $8 00

Consisting of W4390 on mahogany stand; binding posts are mounted upon hard rubber, and a graphite resistance of about 10000 ohms is placed in the circuit and under the base, so as to prevent any accidental short circuiting.

W4394. **Glass Cell, Etc.**..$10 00

Same as W4393, but with permanently mounted thermometer.

W4395. **Mounted Glass Cell** $9 00

W4390 with neat metal case for permanent keeping. With graphite resistance of 10000 ohms to prevent accidental short circuiting. Two milled screws allow the metal case to be lifted from the base.

W4396. **Mounted Glass Cell**.... ....$11 00

W4395 with addition of thermometer.

THE CARHART U. S. STANDARD CELL.

As already stated, the cell prescribed by the National Academy of Sciences, and made legal by Congress, is not portable and must be set up by the user. For the convenience of those not caring to set up their own cells, as well as for those requiring a strictly portable cell, to be moved from one place to another at will and subjected to more or less rough treatment, Prof. Carhart has devised a standard cell, having exactly the same E. M F., and temperature co-efficient as the Congress cell and at the same time perfectly portable. The cell is made up in the "test tube" form common to Professor Carhart's well-known "Carhart-Clark" and "One-Volt" cells.

W4397. **Carhart's U. S. Standard Cell** $12 50

Single cell in brass case with hole for thermometer and special graphite resistance to prevent "short circuiting."

W4398. **Carhart's U. S. Standard Cell** $22 50

Same as W4397. Two cells in one case with separate terminals; can be used singly, in series or in multiple.

W4400. **Carhart's U. S. Standard Cell** $15 00

Single cell, mounted in circular brass case, with rubber top. Has special thermometer permanently mounted upon the rubber top of the case. A graphite resistance of 10000 ohms is placed in series with the cell (inside the case) to prevent any dangerous short circuiting. With certificate of accuracy from Prof. Carhart.

W4401. **Carhart's U. S. Standard Cell**...................$27 50

Same as W4400, but two cells mounted in one case. With certificate of accuracy from Prof. Carhart.

THE CARHART-CLARK STANDARD CELL.

The Carhart-Clark Cell is a modified form of a cell suggested by **Latimer Clark** in 1873. It is made up in a small test tube measuring about $\frac{5}{8}''$ in diameter and about $2''$ in length. The general construction of the Clark Cell is shown by Fig. 71. As now made, **Prof. Carhart** saturates the $Zn\ SO_4$ at $0°$ C.; he also employs various porous separators between the constituents in order to gain greater portability. The seal at the top covers everything hermetically.

The E. M. F. of the Carhart Clark cell is given by the equation:

$$E = 1.441 \left\{ 1 - 0.00039\ (t°C - 15) \right\} \dots\dots\dots\dots\dots (11)$$

which indicates that at $15°$ C. the E. M. F. of the cell is 1.441 volt, and that for every degree C. above or below $15°$ C. a deduction or addition of .00039 volt has to be made.

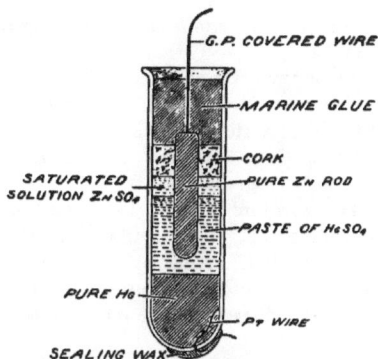

Fig. 71. CLARK CELL.

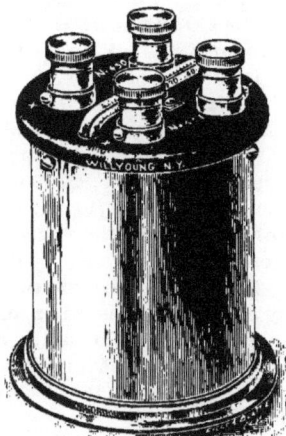

Fig. 72. W4408.

W4405. **Carhart-Clark Standard Cell**......... ....$12 50

Single cell in brass case with hole for thermometer, and special graphite resistance to prevent short circuiting.

W4406. **Carhart-Clark Standard Cell**.....................$22 50

Similar to W4405. Two cells in one case, with separate terminals; can be used singly, in series or in multiple.

W4407. **Carhart-Clark Standard Cell**.....................$15 00

Same as W4405, but with addition of permanently attached centigrade thermometer.

W4408. **Carhart-Clark Standard Cell**................ ....$27 50

Same as W4406, but with addition of permanently attached centigrade thermometer.

W4410. **Set of Seven Carhart-Clark Cells**...............$75 00

Giving 10.08 volts at $15°$ C, mounted in case with graphite resistance and thermometer. Cells can be used individually or in series.

SELF AND MUTUAL INDUCTION.

Definition.—When a current is started in a coil a small time must elapse before it rises to its final and constant value. During this time the rate of increase of the current is not uniform. Nevertheless the whole time required may be imagined as split up into a large number of equal parts, n, any one of which is so small that the rate of increase is practi cally constant throughout the interval. Now if r represents the *rate of increase* during the first of these intervals we know that this will produce an E. M. F., e, and hence we may write

$$e = Lr, \tag{12}$$

where L is the *Coefficient of Self-Induction*. The E. M. F. so produced is, of course, a *back* or *counter* E. M. F. in opposition to the current pro ducing it. We have, thence, the **Definition:**

The **Coefficient of Self-Induction** *in any circuit is the ratio of the counter E. M. F. to the* **time-rate** *of change of current in the circuit.*

It is easy to show that it may also be defined as *the quantity of* induced or extra electricity which is made to circulate in a circuit of unit resistance by making or breaking unit current.

The Henry.—This is the practical or "working" unit of S. I., and corresponds to 10^9 C. G. S. units, *i.e.*, to the cutting of 10^9 magnetic lines, when unit current (10 amperes) is sent through the circuit. This unit is also often called the "secohm," this being the name first applied to it before action was taken by the International Congress.

Mutual Induction.—Suppose two circuits placed close together; any current established in one will induce an E. M. F. in the other, due to the cutting by the latter of the magnetic field set up by the former. The total number of lines cut by the *second* circuit when unit current is started in the *first* circuit, is defined as **The Coefficient of Mutual Induction, M.**

Relation of Capacity and S. I.—Ohm's Law $C = \dfrac{E}{R}$, applies only to steady currents. For alternating or varying currents the expres sion for current may be shown to be

$$C = \sqrt{\dfrac{E}{R^2 + (L\omega - \dfrac{1}{F\omega})^2}} \quad \dots\dots\dots\dots\dots\dots (13)$$

Where L is the coefficient of S. I , F the capacity, and ω the "angular velocity" $= 2 \pi n$ (n being the number of full periods per second). From the second term it appears that if $L = \dfrac{1}{F}$ the equation (13) will reduce to $C = \dfrac{E}{R}$ *i.e. in circuits containing* **both S. I.** *and* **Capacity** the one *may be made to annul the other by a proper choice of values.*

Impedance.—In any circuit containing S. I. only the equation (13) becomes

$$C = \sqrt{\dfrac{E}{R^2 + L^2 \omega^2}} \quad \dots\dots\dots\dots\dots\dots\dots (14)$$

The quantity $\sqrt{R^2 + L^2 \omega^2}$ is called the **Impedance.**

Measurement of S. I. and M.—Two general methods are suggested by the preceding equations (13) and (14). (1) Measurement in terms of a Capacity, *i.e.*, a Standard Condenser, basing measurement on equation (13) and finding C, E and R by orthodox methods. (2) By "Impedance" using equation (14), where, also, C, E and R are readily measurable.

We may, also, have **Standards of S. I.**, in terms of which any unknown Coefficients are very simply determined.

REFERENCES.

Methods of measuring **S. I.** and **M** may be found in
> *Henderson*, pages 300–323.
> *Stewart & Gee*. pages 390–397.
> *Carhart & Patterson*, pages 235–274.

A very simple and practical method is also given in *Kempe*, pages 547–550.

Fig. 73. W4420.

Fig. 74. W4425.

SELF-INDUCTION APPARATUS.

W4420. Secohmmeter............................ .Grade B $50 00

An improved form of the well-known instrument of Profs. Ayrton and Perry. The Secohmmeter will: 1. Compare two co-efficients of self-induction; 2. Compare two capacities; 3. Compare two co-efficients of mutual induction; 4. Measure, absolutely, a co-efficient of self-induction; 5. Measure the actual resistance of a polarizable electrolyte, as well as serve for a large number of other determinations. The instrument consists, essentially, of two commutators mounted upon one shaft and allowing both battery and galvanometer of a Wheatstone's bridge arrangement to be commuted at any desired speed from 300 to 6000 per minute and at an adjustable interval as regards themselves. The "self-induction" kicks are thus converted into a steady, uni-directional deflection. The apparatus may be driven by hand or motor as desired. (See *Henderson*, p. 313, etc.; C. & P., p. 231, etc.)

W4421. Secohmmeter............................ .... $100 00

Same as W4420 but direct driven by 110 volt motor. An attachment is provided for varying the speed within wide limits. Cannot be driven by hand.

W4425. Profs. Ayrton and Perry's Variable Standard of Self-Induction............... Grade B $135 00

Reading direct in Milli-Henrys. Range of variations from 3.5 to 35 M. H's. Of mahogany, built up to prevent warping. (See *Henderson*, p. 315; C. & P., 267.)

W4430. Standard of Self-Induction of one fixed value.. ....
Grade A $50 00
" B 40 00

Mounted in case like ordinary resistance; any value desired from 1 to 100 M. H's.

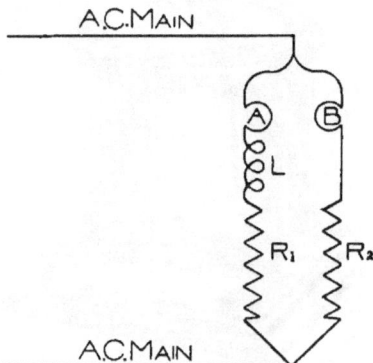

Fig. 75. W4436.

W4431. **Standard of Self-Induction — One
Henry.** ....

                       Grade A $130 00
                       "   B   115 00

Same style of mounting, etc., as W4430.

W4432. **Standard of Self-Induction** ....

                       Grade A $90 00
                       "   B   70 00

With four fixed values of 10, 20, 30 and 40 M. H's, and arranged with plugs like a resistance box. If desired, any or each of these values may be anything from 1 to 100 M. H's without increased cost.

W4433. **Standard of Self-Induction** ....

                       Grade A $125 00
                       "   B   100 00

Same as W4432 but arranged so that when any coil is cut out of circuit it is replaced by one having the same resistance but no S. I.

W4436. **Martienssen's (Pulnj's) Apparatus.**

Price on application.

For measuring Coefficients of self-induction. With this Apparatus L is determined by a species of galvanometric measurement of the same general character as measurement of resistance with the Bridge. The instrument consists essentially of a "Detector" made up of 2 vertical coils, with the planes at right angles to one another. In the centre of these coils is suspended a cylindrical rod of very homogenous copper; a mirror is put on this system so that deflection about the axis may be noted.

Any difference in the phase of two currents passing through these two coils exerts a torque upon the rod and produces deflection.

Referring to the diagram, A and B are the detector coils with inductances A_1 and B_1 respectively. (These must be determined once for all.) L is an inductance to be measured, and R_1 and R_2 adjustable non-inductive resistances; (say boxes ranging from $\frac{1}{10}$ to 10,000 ohms.) The A. C. mains being joined as shown, R_1 and R_2 are adjusted until no deflection is observed.

When this is true we may write,

$$\frac{A+L}{R_1} = \frac{B}{R_2} \quad \ldots \qquad \ldots \ldots \ldots \ldots \ldots \ldots \ldots (15)$$

From which follows at once the value of L.

The range of the instrument may be varied by including the known inductance L_1 with the coil B so as to make L_1+B somewhere nearly equal to $A+L$.

For a full description of this instrument and its use see Wiedemann's Annalen, Jan., '99, " Methode und Instrument zuhr Messung sehr kleiner Inductions Coefficienten," von H. Martienssen.

ELECTRO-DYNAMOMETERS.

Fig. 76. W4450 and W4452.

Fig. 77. W4454-4456.

The Electro-Dynamometer is the *only* **Standard Ammeter for Alternating Currents,** whether they be of large or small value. It contains no iron, and its indications depend only upon the torsion of its control and the current being measured.

The Law of the Electro-Dynamometer.—Let θ be the angle of twist of one coil with reference to the other, and K a constant which, when multiplied into θ, gives current values. The deflecting force at any instant, and hence the twist, will obviously be proportional to the product of the currents in the two coils, or $K\,\theta = C'_1 \times C'_2$, which becomes $K\,\theta = C^2$, when the same current passes through both coils in series. But this is at a given instant. In actual use the coil will tend to take a position dependent upon the *mean* or *average* value of the deflecting forces, or

$$\frac{K\,(\theta_1 + \theta_2 + \theta_3 + \theta_4, \text{etc.})}{m} = \frac{C'^2_1 + C'^2_2 + C^2_3 + C'^2_4, \text{etc.}}{m}$$

from which $\qquad\qquad C_m = K\sqrt{\theta}\,\ldots\ldots\ldots\ldots\ldots\ldots\ldots(16)$

When, therefore, we measure an A. C. by means of the Electro-Dynamometer, we really find the direct current which will require the same twist of the spring, and this value will be equal, numerically, to the *"square root of the sum of the mean squares"* of the current values at successive instants of time.

The Electro-Dynamometer is equally useful for direct current circuits. C_1, C_2, C_3, etc., will then all be equal, and the formula $C = K\sqrt{\theta}$ is immediately obtained.

The Great Merit of the Electro-Dynamometer is that the **constant,** K, may be obtained *at any time* by taking one reading only with a direct current in series with a D. C. Ammeter whose indications are exactly known. Or a voltmeter may be used.

W4450   **Electro-Dynamometer**............................ $60 00
    Range 0.02 to 2 amperes.

W4452.   **Electro-Dynamometer**...................... .... 66 00
    Range 0.04 to 5 amperes.

W4454.   **Electro-Dynamometer**...................... .... 44 00
    Range 0.2 to 20 amperes.

W4456.   **Electro-Dynamometer**............................ 44 00
    Range 1 to 60 amperes.

W4460.   **Electro-Dynamometer**............................ 55 00
    Range 5 to 200 amperes.

W4462.   **Electro-Dynamometer**...................... .... 100 00
    Range 10 to 500 amperes.

NOTE.—The Electro-Dynamometers W4450–4462 are of the type generally known as that of "Siemen's"; they are much used as **Standards of Current** in commercial alternating work. W4450-52 are enclosed, and the moving systems, which are very light, are hung on jewels. In the other instruments W4454-62 the systems are heavier and are suspended by fibres.

PRECISION ELECTRO-DYNAMOMETERS.

For very exact work and in scientific investigation electro-dynamometers are made reflecting just as are galvanometers. The several instruments listed below have been carefully worked out. The frames and mountings are chiefly made up of hard rubber, the little metal employed therein being carefully selected for non-magnetic properties and of high resistivity; as an additional safeguard against such parts becoming the seat of inductive disturbances they are split and segmented wherever possible.

W4465. Precision Electro-Dynamometer.............. $150 00

For measuring alternating currents from 0.0001 to 50 amperes, and direct currents from 0.000001 to 50 amperes.

There are two pair of coils in this instrument, one small pair of fine wire closely surrounding the moving system, and a second pair, made of heavy flat copper strip, outside of these; these latter coils are so held as ventilate freely and are thus kept cool. The moving system is small and light, and by means of a clamping vane is made quite dead-beat.

W4466. Precision Electro-Dynamometer.............. $90 00

Similar to W4465, but measuring to a maximum of but 5 amperes instead of 50.

W4467. Telescope and Scale Attachment............. $25 00

For W4465 and W4466. Arranged to place directly upon the instrument. Scale, 50 cms. long in mm. divisions; zero at end or centre, as ordered.

W4468. Wall Bracket for W4465 and W4466....... $10 00

The instruments are set directly upon this bracket and held firmly by a special clamp which forms part of the bracket.

Special Electro-Dynamometers of higher or lower range will be quoted on application.

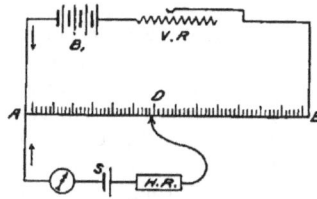

Fig. 78.

POTENTIOMETERS AND THE POTENTIOMETER METHOD.

The Potentiometer Method is, we have always held. the most simple, convenient, and reliable method of obtaining very exact measures of E. M. F. and current in existence. Primarily an E. M. F. is measured by balancing it (or a known percentage of it) against the E. M. F. of a Standard Cell of known value or against a definite proportional part of this known value. To measure Current the E. M. F. (though a Standard Resistance) due to the unknown current is determined, whence, of course,

$$C = \frac{E}{R}.$$

We outline these two applications very briefly below.

To Measure E. M. F.—See fig. 78. S. is a Standard Cell. H R a high resistance (to prevent polarization and accidental short circuit) in series with S, V R an adjustable resistance and B a storage battery of E. M. F. greater than S. The wire. A B, is divided into 1000 parts. Set Slider to the value of the Standard Cell—144.1 (read 1.441 volts) and adjust V R till galvanometer remains at zero ; the E. M. F. along A B will then be as per the scale values, and to determine any unknown E. M. F. it need merely be substituted for S, the Slider moved till galvanometer is zero, and the value read off.

The "Tap Off" Box.—If the unknown E. M. F. is higher than the total of A B we may use an extra resistance box whose total has leads tapped off for $\frac{1}{1000}$, $\frac{1}{100}$, $\frac{1}{10}$, etc. of the whole. Applying the unknown to the main terminals leads are carried from the "taps" to A B and a known percentage of the whole thus measured. Such an extra resistance box is known as a "tap off" box. Any of the Ayrton shunts, W4200 to W4202A, serve admirably for this purpose, the "six-stepped" ones particularly.

Merits of the Potentiometer Method.—The Potentiometer method depends only upon a Standard Cell and a Standard Resistance, the behavior of both of which has been studied and observed under all sorts of conditions for a number of years. As a result of this study, it has been conclusively shown that a Resistance Standard made properly, and as many makers know how to make, and with reasonably careful use should not change by more than a few 100ths per cent. in years. Good

Standard Cells, also such as may readily be obtained upon the market, are *known not* to change by more than a maximum of five or six parts in 10,000 in, say, four or five years. Furthermore, by keeping an extra Standard Cell as an ultimate "Master Standard," an occasional comparison will at once show whether *either* has changed since the *same* change in both would be extremely improbable. Similarly, the resistances of the Potentiometer are generally so grouped that inter-comparison is easy, so that change in them is quickly noted just as for the Standard Cells.

Another merit of this method is that it is a *zero* method; the Standard Cell is *not* developing current, and the galvanometer *deflection* values need *not* be known; provided only that the galvanometer is *sensitive* at the zero point not even the value of this sensitiveness need be known.

WILLYOUNG DIRECT READING POTENTIOMETER.

W4520. **Willyoung Direct Reading Potentiometer** ..$175 00

The plan of connections, etc., is shown in the diagrammatic view of fig. 79. There are three sets of resistances in the main circuit, the first set reading 0.1, 0 2, 0.3, etc., to and including 1.4; the second 0.01, 0.02, 0.03, etc., to and including 0.09; and the third 0.001, 0.002, 0.003, etc., to and including 0.009. There are two plugs in the first set, one in the second, and one in the third, *thus making but four contact resistances requiring change during the operation of the instrument. Two of these are in the derived circuit and hence can introduce no error*, while the remaining two may be very positively counted on to produce no sensible error (being well made plug contacts) in the main potentiomer circuit of 200 ohms.

In the working of the instrument, switch A is set to point 1, thus opposing the E. M. F. of the standard cell to that of the storage cell. The plugs are then set to the E. M. F. of the standard cell and "Regulator Quick" and "Regulator Slow" altered until on closure of the double contact key, there is no deflection of the galvanometer. Preliminary trials may be made with the galvanometer shunted, a special switch being provided for this purpose. Balance being secured, turn A to point 5 and bring the unknown E. M. F. to be measured to the posts 1 and 5. Alter position of plugs until balance is again obtained, when the readings of the plug positions will give the E. M. F. Should the unknown E. M. F. be more than 1.5 and less than 15 volts, bring A to point 4 and proceed as before. The plug readings must now be multiplied by 10 to give the E.M.F. Point 3 is used for E. M. F.'s between 15 and 150 volts, and point 2 for E. M. F.'s between 150 and 1500 volts.

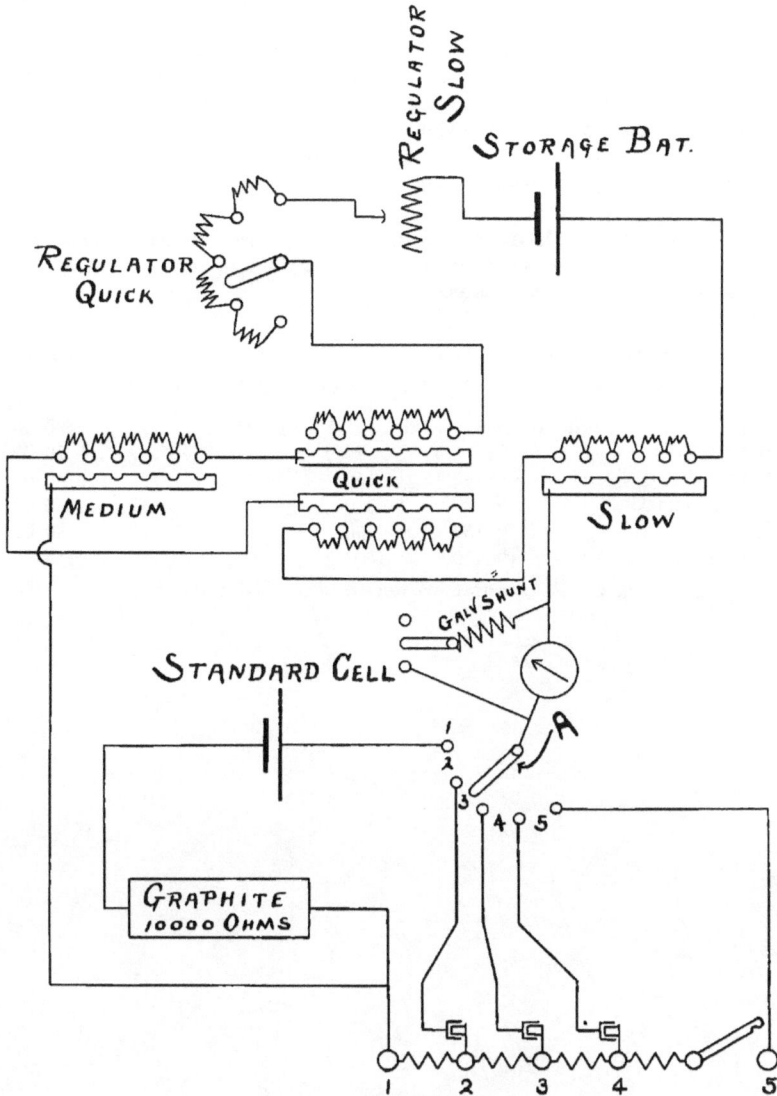

Fig. 79. DIAGRAM OF W4520.

Recent Changes in Design.—The high resistance, "B," is 100,000 ohms; in our present and recently revised type of instrument this "tap-off" resistance is contained in a separate "tap-off" box with stud-tipped flexibles for connecting to the main box. Any heat developed by high E. M. Fs. is thus incapable of affecting the resistances in the Potentiometer proper. Battery and Galvanometer keys, also, are now made separate, the former being a neat and small, but effective *switch*, giving good generous surface at contact. The various individual resistances, also, of the Potentiometer have been arranged so that they may be isolated and separately measured, thus enabling any possible change in their values to be instantly detected.

Detailed working drawings of this instrument sent to responsible parties; there has not been time to secure a cut of the revised model for this catalogue.

STANDARD LOW RESISTANCES FOR USE WITH THE WILLYOUNG D. R. POTENTIOMETER.

W4530.	1.	Ohm, to carry 0.15 to		1.5 amperes			...:........	$20 00
W4531.	0.1	"	" " 1.5	" 15	"		....	20 00
W4532.	0.01	"	" " 15	" 150	"		....	45 00
W4533.	0.005	"	" " 30	" 300	"		....	55 00
W4534.	0.001	"	" " 150	" 750	"		....	90 00
W4535.	0.0005	"	" " 150	" 1500	"		....	135 00

These Standard Resistances are accurate to $\frac{1}{25}$ of one per cent.

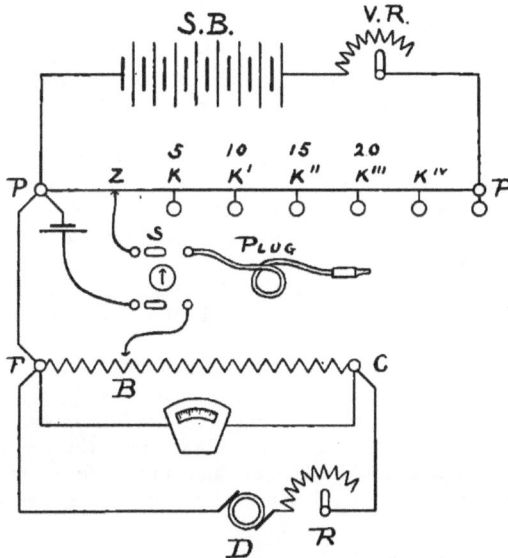

Fig. 80. DIAGRAM OF CHECKING POTENTIOMETER.

THE CHECKING POTENTIOMETER.

The requirements of modern Central Stations, Municipal Laboratories, City Police and Fire Department Bureaus, etc., are such as to necessitate the employment of very large numbers of Portable and Switchboard Volt and Ammeters. Some way of quickly "checking over" the scales of such instruments from time to time is obviously of prime importance. The "Checking" Potentiometer has been designed for this purpose. It is a Potentiometer pure and simple, made to do exactly what is wanted (and only this) in the simplest possible way.

It is assumed that the exact value of E. M. F. or Current for every possible reading of the instrument being checked is not necessary, but only the values for a certain number of scale readings uniformly distributed over the scale. The instrument is, hence, so arranged.

Fig. 80 shows the 150 volt instrument in diagram. P P is the main Potentiometer wire, S B an 8-cell storage battery, and S C a standard Carhart-Clark Cell. Sections P K, K K', K' K", etc., are coils of equal resistance and each $\frac{1}{30}$ of the whole resistance P P; at each point K, K', K", etc., is a plughole. A B C is an outside resistance (15,000 ohms); F B steps off exactly $\frac{1}{10}$ of this (1,500 ohms). S is a double pole switch by which either the Stnd Cell circuit or that from F B may be placed in series with the galvanometer and in opposition to the E. M. F. due to S B.

The operation of the instrument is simple. Throw S to the left, thus putting the Stnd Cell in service and adjust V R until the galvanometer does not deflect. A shunt (not shown in the figure) around the galvanometer may be removed for the final balancing. The position of Z, permanently fixed inside of the case, is such that. at 20°C, exactly 15 volts difference of E. M. F. exists between P and P. Now throw S to the right, thus "tapping-off" $\frac{1}{10}$ the E. M. F. between F and C, insert plug into any hole and quickly adjust the E. M. F. of D (the outside E. M. F. which operates the Voltmeter to be checked) until the galvanometer is again at zero. The actual voltage upon the Voltmeter is then 10 times that of F B (= 10 times that between P and plug hole in use); hence. if the E. M. F. to any plug hole is marked at the hole with its value multiplied by 10 the instrument becomes direct reading.

For temperatures above or below 20°C add or subtract respectively $\frac{1}{1000}$ for each 5°. This correction is so small that it need not be considered in the general case.

The whole instrument is compact and enclosed in a polished mahogany case, with handles, lock and key. There is nothing to get out of order and nothing exposed. There is nothing to manipulate except switch S, the plug cord, the galvanometer and shunt key, and the resistance V R. Nothing more practical could be desired and any "handy" man can learn its use in a few moments Brief explicit directions accompany each instrument.

PRICE LIST.

W4540. **Checking Potentiometer,** 0 to 600 volts by 25 volt
steps... $140 00
Complete as described, except Standard Cell and Storage Batteries.

W4541. **Checking Potentiometer,** 0 to 300 volts by 10 volt
steps..................................... .... 125 00
Otherwise like W4540.

W4542. **Checking Potentiometer,** 0 to 150 volts by 5 volt
steps........ .... 100 00
Otherwise like W4540.

W4546. **Portable Storage Battery** of 8 cells for use with
the Checking Potentiometer, in oak case with handle.. 35 00

For Standard Carhart-Clark Cells for use with the Potentiometer see page 67. We advise purchase of two cells so that any possible change may at any time be detected by simply trying one against the other.

Current Measurements with the Checking Potentiometer may be made by use of the resistances W4530 to W4535.

THE WILLYOUNG CONDUCTIVITY BRIDGE.

For measuring the conductivity of any and all conductors of uniform cross section in $2\frac{1}{2}$ foot lengths from the equivalent of $\frac{1}{2}$ square copper up to the smallest section of any alloy. The most simple, practical, reliable and accurate means of measuring Conductivity which has ever been suggested. Adopted and in use by the following manufacturers and institutions:

THE BRIDGEPORT BRASS CO.,   THE GUGGENHEIM SMELTING CO.,
THE BRIDGEPORT COPPER CO.,   THE WACLARK WIRE CO.,
THE SEYMOUR MFG. CO.,   THE LAKE SUPERIOR MINING & SMELT-
THE COE BRASS CO.,   ING CO.,
ROEBLINGS SONS CO —*have four*   THE STANLEY ELECTRIC CO.,
    in daily use,   MICHIGAN SCHOOL OF MINES.

This instrument takes advantage of the Carey-Foster method of using the slide bridge, now almost universally used by the leading laboratories and manufacturers of the world in the exact comparison and measurement of standards.

In the Carey-Foster method of using the bridge, theory shows that if we have two resistances in the two outside gaps of a four-gap bridge and we exchange these two resistances, obtaining a balance upon the bridge wire each time, then the actual resistance of the bridge wire moved over from the one position of balance to the other, is the exact *difference* of resistance between the two coils. Contact resistance, resistance of connections, etc,, all eliminate themselves and are of no effect.

In the conductivity bridge this idea is reversed by having a known and accurately predetermined *difference* between the two coils, and *using as a bridge wire* the length of copper whose conductivity is desired. Variations of conductivity are thus shown upon different samples of the same cross sectioned wire by varying distances between balance positions.

The measurement is entirely independent of temperature, no matter at what temperature the bridge be used, and is always in terms of the temperature, usually 70° F, at which the bridge was standardized.

DESCRIPTION.

This instrument was first listed by us in March, 1897. Since that time continuous work with the bridge in various prominent manufactories has confirmed our belief in the immense superiority of our method and apparatus for conductivity measurements over all others. In the factory of the Roeblings' at Trenton, an average of forty samples a day is regularly measured by one man, while on some days as many as one hundred and ten to fifteen have been measured. Careful checking of this bridge from time to time during the past year has shown that despite the continuous and heavy work that it is doing, no change in the constants of the bridge has taken place in this time.

To protect the instrument from air currents and radiation and thus secure evenness of temperature throughout the whole bridge is permanently mounted within a polished oak case. The milled head operating slider is passed through the front so that the bridge may be operated with the case closed. The lid of the case is so designed as to leave a low front and ready access to the bridge proper when case is opened. Key and binding posts are mounted on top of the lid and joined to the bridge proper by interior flexibles.

Bridge slider moves over a silvered and machine divided brass scale, and has both coarse and fine (rack and pinion) adjustments. Metal work is of very massive copper, and woodwork of massive well dried and braced mahogany.

Below is given an extract from the Sheet of Directions sent out with the bridge; careful reading of these directions with reference to the diagram will make the theory and *modus operandi* clear.

EXTRACT FROM DIRECTIONS.

To use the Bridge ; join the two posts G to the terminals of a suitable galvanometer and posts B similarly to a source of current; a single cell of storage battery, Daniel or other fairly low internal resistance battery will answer. A resistance had best be arranged in series with the cell so that not over about one ampere at the most can ever flow into the bridge. If this precaution is not observed dangerous heating of the standard resistances of the bridge may take place.

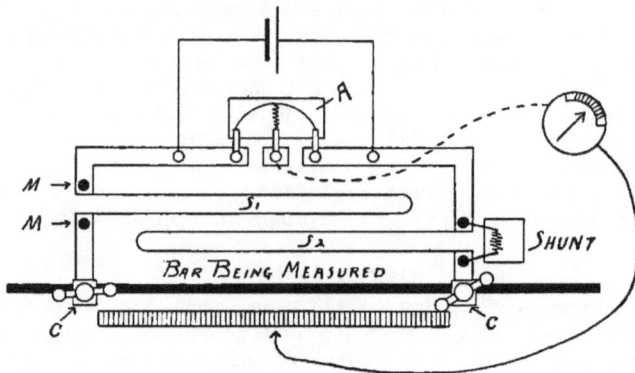

Fig. 81.

Now cut the conductor whose conductivity is desired to a length suitable for use, place it in the jaws of the bridge (see Fig. 81) and clamp down tightly. Place the proper shunt in the mercury cups at either end of the bridge and move slider to balance (no deflection of the galvanometer is obtained). Wait a few moments and carefully get the last reading again, in this way assuring a uniform temperature condition. Now note the reading, move the slider over to approximately the same distance the other side of centre, place the shunt in other end of bridge, close the case and get a new balance.

Suppose the two balances to be B^1 and B^2 respectively, then B^1-B^2 $=L'$ Having thus obtained L' we now obtain the conductivity C in either of the two ways

(*A*) By knowing weight in length where C=Conductivity

Using Formula

$$C=\frac{BLL'K}{WR} \cdot 100$$

B=Resistance of one metergram of copper at 70°F (Matthiessen Standard)=0.1560

W=Weight

L=Length weighed

K=A constant=0.0929

When W is in grams and L and L' in feet. W=Weight in grams of sample

R=Resistance of L'=shunt value used

The scale is graduated in $\frac{feet}{100}$ the vernier thus reading to $\frac{feet}{1000}$.

It is most convenient to adopt a standard length for L, which thus becomes a constant; using logarithms this formula is very simply applied. The formula

$C=\frac{BLL'K}{WR}$ may also be written.

$C=\frac{K''L'}{W}$ Where K'' is a new constant made up of the old constants B, L, K, and R; in this form the slide rule may advantageously be used.

(*B*) By knowing the diameter using the formula

$$C=\frac{AL'}{D^2R} \cdot 100$$

Where A=Resistance of one mil-foot of copper at 70°F (*Matthiessen Standard*)=10.397.

D=Diameter of sample in mils.

L' and R having the same significance as before.

Results—will be accurate to within $\frac{1}{10}\%$ and are in terms of the percentage conductivity at 70°F.

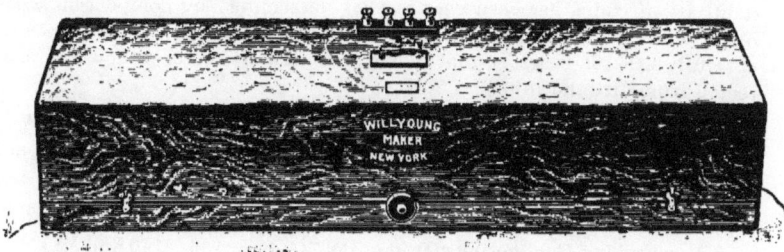

Fig. 82. W4550-W4551. CLOSED.

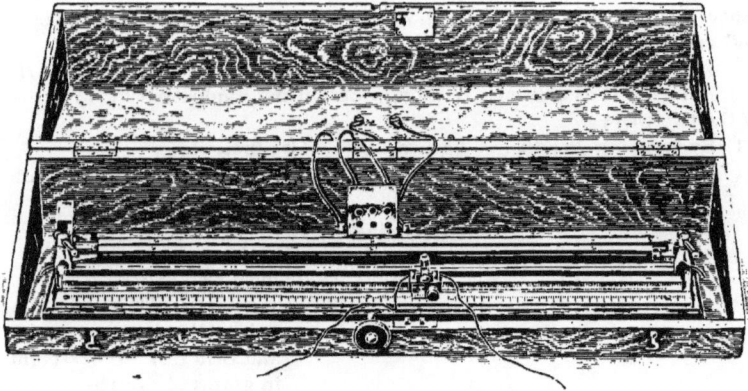

Fig. 83. W4550-W4551. OPEN.

W4550. Willyoung Conductivity Bridge.............. $225 00

For measuring the conductivity of copper wires and rods in 2½ feet lengths and from No. 6 to No. 12 B. and S. inclusive. As used by the Bridgeport Copper Co.; John A. Roebling's Sons Co.; Seymour Manufacturing Co.; Coe Brass Manufacturing Co ; Bridgeport Brass Co.; Waclark Wire Co.; Guggenheim Smelting Co.; Michigan School of Mines; Pittsburg Reduction Co.; Stanley Electric Co.

W4551. Willyoung Conductivity Bridge.............. $250.00

Similar to W4550, but adapted to measure wires from No. 4 to No. 16 B & S inclusive.

STANDARD BARS FOR CONDUCTIVITY BRIDGES.

In the continuous use of the Conductivity Bridge there is always a chance that the resistances or shunts may change slightly. It is, therefore, advisable to check the constants of the shunts occasionally. For this purpose we provide standard samples of carefully annealed copper, the conductivity of which has been very exactly determined by comparison with our "Master" Standard, which is Cavendish certified. These bars are carefully fitted in cases, so that, when not actually in use, they are protected from dirt and injury.

W4560. Set of Four Standard Rods $10 00

For checking Conductivity Bridge B5370—Nos. 6, 8, 10 and 12 in containing case.

W4561. Set of Five Standard Rods........................12 50

For checking Conductivity Bridge B5371—Nos. 4, 6, 8, 10 and 12.

WILLYOUNG'S SPECIFIC RESISTANCE
AND TEMPERATURE COEFFICIENT APPARATUS.

W4562. This apparatus has recently been devised by Mr. Willyoung for the purpose of quickly determining with accuracy the specific resistance of various gauges of wire (or preferably, in practical work their resistance per unit length, which is the same thing), and the "temperature coefficient" or per cent. rise of resistance with rise of temperature.

The apparatus is essentially a miniature **Willyoung Conductivity Bridge** (see page 81, etc.), with some small differences of detail. A 6″ piece of the wire to be tested is clamped into vises arranged to receive it. The two standard resistances are wound upon the same small spool and mounted upon a little ebonite platform in the centre of the apparatus. The shunt required at any particular time is also mounted upon a spool and slips into sockets attached to a rocking device in the just-mentioned platform; when the rocker is thrown one way the shunt parallels one of the resistance standards, and when thrown the other way parallels the other. Various shunts are provided according as the resistance of the test wires is more or less, and may be instantly interchanged in the rocker at will.

The whole instrument is enclosed in a case, upon the inside of which and surrounding the apparatus is a heating coil. For temperature coefficient work the temperature of the apparatus and test wire may be held at any desired point by passing a suitable current through this heating coil.

In operation the wire is clamped in place and the glass lid closed, a proper shunt being inserted first. The slider is pulled along by the outside string until balance is secured. The scale is then read through the glass. Then tip the rocker to the other side by the outer finger provided for the purpose, balance again and read. The length of wire between the two balances has the same resistance as that of the shunt.

For temperature coefficients the preceding operation is gone through with at two different temperatures.

The whole apparatus is compact, measuring but 12x10x5″ deep (over heating box), and is well finished in mahogany.

Working drawings, etc., will be furnished those interested.

Price on application.

INSULATION AND CAPACITY TESTS, BREAK DOWN TESTS, LOCALIZATION OF FAULTS, ETC.

Importance of Tests.—Telegraph and Telephone lines, Power lines, etc., are costly of erection both in time and money. Any interruption in this service is an immense inconvenience, and often loss, to the user who has learned to rely upon it and to the owner whose revenue is thus stopped. There is also the actual cost of repairs, in the case of underground and submarine cables particularly, a very costly operation. The importance of constant and careful tests during manufacture and installation of cable or line so as to catch incipient defects before they become serious or to correct serious ones before an entire *system* is made dependent upon it should, therefore, be obvious. The expense of the necessary testing apparatus is absurdly small compared with the financial (and often to human life) safeguard afforded.

Properties of a Transmission Line.—The three principal characteristics of every line affecting its efficiency are: Conductor Resistance (C. R.), Insulation Resistance (I. R.), and Capacity (F.). On alternating lines there is another, viz., inductance, but this, given the material, dimensions of and location of the line, is generally a constant factor and need not be considered.

Conductor Resistance.—This means an ordinary measurement of resistance by any desired method. If the conductor is of very large cross-section the Conductivity Bridge (W4550, etc.) may be used; if not, then a simple Wheatstone Bridge measurement will usually be adequate. C. R. is practically constant, save for its variation with temperature, for any given line unless breaks or grounds, partial or complete, take place (see remarks on Faults and Fault Testing).

Insulation Resistance.—The insulation resistance of a line strictly considered consists of two parts; the actual resistance of the insulating coating in which the line may be embedded, or the dielectric resistance; and the resistance to leakage afforded by the insulators, poles, troughs, etc., by which the line is actually supported. The former only exists, of course, in *covered* wires or cables and is an actual resistance through the *substance* of the insulator. The latter exists with both cables and bare conductors, but makes up an appreciable percentage of the whole I. R. only in the case of overhead or "strung" lines, and it is, further, almost entirely a surface leakage. It is, hence, very dependent upon atmospheric conditions, being very low in wet weather and very high in dry. Moisture after drought, the insulator, etc., having become coated with dust, causes an exceedingly low value, which rapidly rises so soon as rain has effected thorough washing

Cable Insulation.—The insulation resistance of a covered cable is mainly a true dielectric resistance (D. R.). It depends (1) upon the specific resistance of the insulation and (2) on the dimensions of the same with reference to the conductor. It may be expressed by the equation

$$R = \frac{\rho \ log \ \frac{D}{\varepsilon \ d}}{2\pi \ l} = \frac{\rho \ log \ \frac{D}{d}}{2.728 \ l} \quad \dots\dots\dots\dots\dots\dots(17)$$

where ρ is the specific resistance, D the diameter of the *covered* conductor, d the diameter of the bare conductor, and l the length.

With constant dimensions the D. R. depends upon (1) specific resistance of the insulation, (2) its temperature, and (3) the pressure (lbs. per sq. in.) to which it is subjected. The chief causes of deterioration are exposure to air, light and heat.

D. R. diminishes, in general, with rise of temperature and *vice versa*. For gutta percha the D. R., roughly, doubles itself for every fall of 9° F. above 60° F.; for india rubber there is a doubling for every 27° F.

D. R. for gutta-percha increases about $\frac{1}{30}\%$ for each additional lb. per sq. in. of pressure. In submarine cables, therefore, the D. R. should be much higher after than before laying.

Capacity of Cables.—The electrostatic capacity of overhead lines need not, in general, be considered except in case of telephone conductors where even the small capacities met with tend to spoil articulation on long lines. On cables, however, the capacity may amount to as much as $2\frac{1}{2}$ M. F. per mile. It thus becomes a very measureable quantity.

Value of Tests.—During manufacture both D. R. and capacity tests are useful in telling us whether or not the cable is running uniformly. The quantitative value of D. R., also, must be known in order that we may be *sure* the cable will stand the E. M. F.'s to which it is to be subjected in practice. During laying continued D. R. tests tell us immediately of the presence of "grounds" either serious or incipient. And after laying we may, by making systematic and regular tests of D. R. and capacity, detect *weakness* at any point immediately it is begun and long before eventual breakdown.

Methods for Tests.—Except in certain special cases or where the cable is exceedingly short, what is known as the **Direct Deflection Method** will be found the simplest and most suitable for Insulation tests and, for **Capacity tests**, the direct comparison of the "throw" given by the cable with that given by a standard condenser.

Direct Deflection Method.—In this method of measuring exceedingly high resistances what we do, essentially, is to place the resistance in series with a sufficiently sensitive galvanometer and a sufficiently high E. M. F.; the current thus sent through will produce a certain deflection. Then take a definite fraction of this E. M. F., and get another deflection through a known resistance of, say, 50,000 or 100,000 ohms; for this the galvanometer will possibly require shunting. If now we know the law of deflection of the instrument and the ratio of the E. M. F.'s it is clear that we also have at once the *ratio* of the resistances involved; hence the *unknown* high resistance.

More exactly see fig, 84. The cable is grounded by its sheathing and one terminal of B is also grounded as shown. Let the number of cells used be N, and the E. M. F. of each cell be E. Then we may write

$$KD_1 = \frac{N_1 E}{X+G+B} \quad \cdots\cdots\cdots\cdots\cdots\cdots\cdots\cdots (18)$$

when D_1 is the deflection, K the constant necessary to convert deflections into current, G the galvanometer resistance, and B the battery resistance

Now substitute for the cable a known standard resistance R, shunt the galvanometer with a shunt S, and use N_2 cells of battery so as to get a suitable deflection, D_2.

We may write

$$\frac{G+S}{S}KD_2 = \frac{N_2 E}{\frac{GS}{R+\frac{GS}{G+S}+B}} \quad \text{or} \quad KD_2 = \frac{N_2 E}{\frac{GS}{R+\frac{GS}{G+S}+B}} \cdot \frac{S}{G+S} \cdots\cdots\cdots\cdots (19)$$

Dividing, now, both numbers of equation (19) by corresponding numbers of (18), we have

$$\frac{KD_2}{KD_1} = \frac{\dfrac{N_2 E}{\frac{GS}{R+\frac{GS}{G+S}+B}} \cdot \frac{S}{G+S}}{\dfrac{N_1 E}{X+G+B}}$$

from which, solving for X,

$$X = \frac{D_2}{D_1} \cdot \frac{N_1}{N_2} \cdot \left\{ \frac{G+S}{S}(R+B)+G \right\} - (G+B) \cdots\cdots\cdots\cdots 20)$$

Generally G+B is small compared with X, i. e., it may be a few thousand ohms compared with, say, 50 megohms for X; B, also, will be quite small relatively to R. Hence, (20) reduces to

$$X = \frac{D_2}{D_1} \cdot \frac{N_1}{N_2} \cdot \left\{ \frac{G+S}{S} \cdot R+G \right\} \cdots\cdots\cdots\cdots\cdots\cdots (21)$$

Suppose R to be a 100,000 ohm coil and the shunt to be the 1,000 value, i. e., $\frac{G+S}{S}=1,000$, (21) becomes

$$X = \frac{D_2}{D_1} \cdot \frac{N_1}{N_2} 100 \text{ megohms} \cdots\cdots\cdots\cdots\cdots\cdots (22)$$

G, the galvanometer resistance, being neglected.

In applying the "direct deflection" method there is a first rush of current into the cable on closure of the battery circuit. Hence, the galvanometer is short circuited and only opened when the current has settled down to a steady value. If the cable is a good one the deflections will be found to steadily decrease at a continuously decreasing rate, i. e., tending eventually to a constant value. The insulation resistance is. hence, apparently also changing. When but one value is to be recorded it is usual to read at the end of one minute or after "one minute's electrifica-tion."

This change of deflection is due to "absorption." It is probably, really, a kind of electrolytic polarization of the dielectric. In submarine cable work it is customary to take the deflection at the end of each minute for thirty minutes. The cable is then short-circuited for an instant so as to clear it of its excess charge and then put in series with the galvanometer; there will be deflections, constantly decreasing, due to the "soaking out" of the absorbed charge. If the *"discharge"* deflection at the end of any minute be subtracted from the *"charge"* deflection for the end of the corresponding minute the two should differ by the same amount for each set of pairs; this amount will be the deflection corresponding to the *real* insulation resistance.

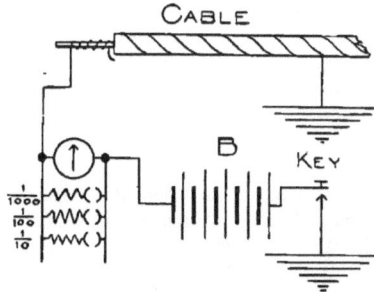

Fig. 81.

Though the actual I. R. of a cable may change from day to day this "constant" relation between the charge and discharge pairs has been found to always exist if the cable is in good condition, while in imperfect cables the relation is very jerky and irregular. A simple taking of the deflections and an inter-comparison of their values is, hence, just as good and better in the case of cables which are tested daily than a working out of the actual value of the I. R.

Capacity of Cable.—Every transmission line is an electrical condenser of which the conductor is one sheet, the ground, water, or various "grounded" surrounding bodies the other sheet, and the insulating covering, air, etc., whatever the case may be, the dielectric. In making capacity tests, therefore, treat the cable exactly as if it were an ordinary condenser connecting to conductor and "ground" respectively.

Choice of Instruments.—For tests during manufacture, etc., a permanent installation is most satisfactory. Several such layouts are given below. The connections are all permanently made by stiff wires carried around on our "H. L. Line Posts." All the necessary changes of connection for the different tests are effected by simple moves of some two or three plugs.

Break Down Tests.—Insulation Tests by Break Down.—It is sometimes desirable to test a cable by increasing E. M. F.'s until it breaks down. To do this, use only a short piece of cable, say two feet long. Have either a *high potential* transformer, or a series of *step up* transformers. Wrap a piece of copper foil about one inch wide around the cable at its middle for one terminal and let the cable core be the other terminal. Suspend the whole arrangement in a tank of resin oil to prevent surface leakage. Place in parallel with this sample a spark gap consisting of two *pointed* and clean steel needles. Run the apparatus with a given gap for about one minute. If no break down occurs, increase the gap by $\frac{1}{16}$ or $\frac{1}{8}$ inch, first shutting off the current, of course, and run again. Continue until break down occurs. According to Steinmetz, the E. M. F. in volts will then be given by

$$\text{E. M. F.} = 10,000 \times \text{gap (in inches)}\dots\dots\dots\dots(23)$$

Specific Resistance, Resistance of Insulators, Resistance of Short Pieces of Cable, etc.--The upper limit of resistance which can be measured by the direct deflection method depends upon (1) the sensitiveness of the galvanometer and (2) the amount of battery used. In practice, using the best obtainable values for both of these probably 100,000 megohms is the upper limit in the laboratory, while in street work from 10,000 to 20,000 megohms may be obtained. The I. R. of short cables as also specific resistances often to be measured are apt to run much higher than these figures. Such exceedingly high resistances may readily be measured, however, by "*leakage*" methods, *i. e.*, they are charged to a given potential as a condenser and their *rate* of loss of charge observed; it is evidently a function of the I. R. The theory of these methods is too long for admission here, but those interested are referred to. *Henderson*, p. 71, etc.; *Kempe*, p. 241, etc.; *Gray, Vol. I*, p. 399.

FAULTS AND FAULT TESTS.

The term **"Fault"** is applied, in electrical phraseology, only to a "*line*" and signifies in a general sense, every variety of damage by which its electrical efficiency is injured. The term "*ground*" is more or less synonomous with "*fault*," but usually implies a relatively low resistance path from line to ground, whereas a fault may be an actual break without connection to earth.

"*Faults*" and "*Grounds*" are of two kinds, *complete* and *partial*. Thus a wire may break and fall with its end in a puddle of water—*very low resistance* and a complete or *dead* ground. Or, the broken end may fall into the cleft of a dry tree—rather high resistance and a "*partial*" ground.

A Fault is shown to exist between transmitting and receiving stations when the proper amount of current cannot be sent to the latter.

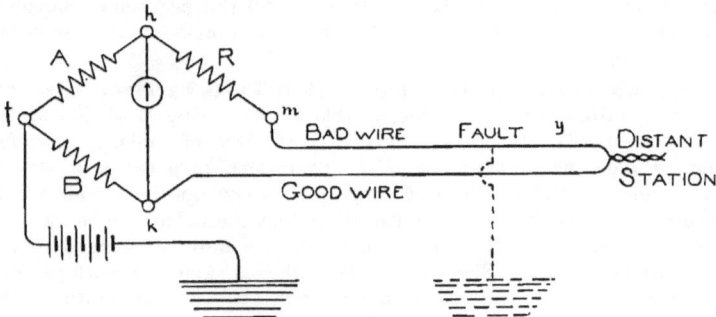

Fig. 85. VARLEY'S LOOP TEST.

The Loop Test. In overhead telegraph and telephone work, faults or grounds are very easily located. If there are intermediate stations they may be cut off one after another until the two stations involved are found. A " Loop Test " may then be employed.

Varley's Loop Test.—See fig. 85. A, B and R are the three arms of a regular Wheatstone Bridge, A and B being the bridge arms proper. The faulty wire is joined to a second good wire, C, at the first station beyond the "fault." The loop is then joined up in place of the usual unknown resistance. If, now, we balance we shall have

$$\frac{A}{B} = \frac{R + X}{C + Y} \qquad \qquad \ldots (24)$$

and, letting $K = X + Y + C$ (the total resistance of the loop which may be obtained from records) from which $C + Y = K - X$, and, substituting this value of $C + Y$ in (24), we may write

$$X = \frac{AK - BR}{A + B} \qquad \qquad \ldots (25)$$

a value for X, which is independent of the resistance of the fault.

Murray's Loop Test.—This is exactly the same as the Varley Test except that the resistance R (fig. 85) is omitted, the loop being taken direct to H and K. For balance we will have

$$\frac{A}{B} = \frac{X}{Y + C}$$

This may also be written

$$\frac{A}{A + B} = \frac{X}{X + Y + C}$$

If the good and bad wires are both of the same gauge and hence of resistance proportional to length for the whole loop we may use a slide wire for $A + B$. This makes the work very easy, thus, if

$$\frac{A}{A + B} = \frac{1}{3}$$

we know that the fault is $\frac{1}{3}$ the way around or, if the distance *between* stations is three miles, $\frac{a}{B} = 2$ miles from the testing station. The " Murray Loop," like the Varley, is also independent of the actual fault resistance.

Faults by Capacity Test.—In case of a total break where the ends remain insulated from the ground a capacity test will locate the fault since the two parts will have capacities proportional to their lengths.

First Method.—Get the capacity of wire on one side of the fault from one end of the cable ; then go to other end and get the capacity of the other piece. Let C and C^1 be these two capacities and X the distance in feet from the first end to the break; then $X : D :: C : C + C^1$ when D is the whole length of cable. Hence

$$X = \frac{CD}{C + C^1} \qquad \qquad \ldots (26)$$

Second Method.—If a good wire is available take a throw, d, from the broken wire, another, d', from the good wire, and a third, d'', from the good wire with the broken wire hitched to it at the far end. The throw from the whole broken wire would be $d'' - d' + d$. Hence, $X : D :: d : d'' - d' + d$, from which

$$X = \frac{dD}{d'' - d' + d} \qquad \qquad \ldots (27)$$

OUTFITS FOR CABLE TESTING.

We have arranged the following outfits, in addition to our Portable Cable Testing Sets W4570–4572. Of these No. 1 Outfit is the simplest arrangement and consists only of those pieces needed for accurate and quick measurement of the Insulation Resistance.

No 2 Outfit is the same as No. 1 Outfit, except that a number of pieces have been added so as to permit of Insulation Tests and Capacity Tests.

No. 4 Outfit covers Insulation Capacity and Resistance Measurements as stated. It is complete and cannot be improved upon.

In any of the preceding outfits other individual pieces may of course be substituted. We shall be glad to make special pieces for such special arrangements as our customers may choose.

With each outfit we supply a scale diagram showing the "lay-out" and full directions as to setting up and operating.

No. 1 OUTFIT.
For Insulation Measurements only.

W4135. Willyoung High Sensibility Galvanometer.
W4139. Tube for same.
W4201. Universal Shunt Box.
W4211. Lamp and Scale.
W5000. 100 Cell Silver Chloride Battery.
W4330. Standard 100,000 Ohm Box (square pattern).
W4364. Willyoung Improved Battery and Short Circuit Key.
W4386. Set of three Cable Posts with Connectors.
W4389. High Insulation Line Posts.

No. 2 OUTFIT.
For Insulation and Capacity Tests.

W4135. Willyoung High Sensibility Galvanometer.
W4139. Tube for same.
W4138.   "   "   (Ballistic).
W4201. Universal Shunt Box.
W4211. Lamp and Scale.
W5000. 100 Cell Silver Chloride Test Battery.
W4330. Standard 100,000 Ohm Box (square pattern).
W4056. Standard Condenser.
W4364. Willyoung Improved Battery and Short Circuit Key.
W4370. Reversing Key.
W4388. Set of five Cable Posts, with Connectors.
       Set of four Terminals on Base.
W4389. High Insulation Line Posts.

No. 3 OUTFIT.
For Insulation and Capacity Tests.

Same as the No. 2 Outfit, but with W4058, Subdivided Condenser, substituted for the plain Condenser W4056.

No. 4 OUTFIT.

For Insulation, Capacity and Resistance Measurements.

W4135. Willyoung High Sensibility Galvanometer.
W4139. Tube for same.
W4138.   "   "   (Ballistic).
W4201. Universal Shunt Box.
W4211. Lamp and Scale.
W5000. 100 Cell Silver Chloride Test Battery.
W4330. Standard 100,000 Ohm Box (square pattern).
W4056. Standard Condenser, ½ M. F.
W4302. Standard Resistance Box and Bridge.
W4364. Willyoung Improved Battery and Short Circuit Key.
W4370. Reversing Key.
W4388. Set of five Cable Posts, with Connectors.
       Set of four Terminals on Base.
W4389. High Insulation Line Posts.

No. 5 OUTFIT.

For Insulation, Capacity and Resistance Measurements.

Same as No. 4 Outfit, but with W4058 Subdivided Condenser substituted for the plain Condenser W4056.

PORTABLE CABLE-TESTING SET.

W4570. **Portable Cable-testing Set** $167 50

For measuring the insulation resistance of underground and other
high grade cables during and after laying, See *Electrical World*, Jan. 2,
1897; *Electrical Engineer*, Dec. 15, 1896; *Electrical Review*, Jan. 13, 1897;
Western Electrician, Dec. 18, 1896.

As used by the Union Traction Co; Central District and Printing Tele-
graph Co,; Hartford Electric Light Co ; St. Anthony Falls Water Power
Co.; Postal Telegraph Cable Co. ; The Valley Telephone Co.; Safety Insu-
lated Wire and Cable Co.: Washburn & Moen Manufacturing Co. ; Central
Union Telephone Co.; Borough of Manhattan Electric Co.; District
Engineer, Washington, D. C.; Tucker Electrical Construction Co , Roeb-
lings Sons Co. (three successive orders), Syracuse Electric Light and
Power Co.

Consists of D'Arsonval galvanometer, with telescope and scale; high
resistance of ten, twenty, thirty and forty thousand ohms; Ayrton Universal Shunt; galvanometer short circuit key, battery make circuit key and
discharge key; and terminals for attachment to cable under test. Several
such terminals are provided so that changes may be made from one cable
to another by mere change of a plug. The galvanometer is hung upon
gimbals and is adjustable in two planes for levelling. Shunts are four in
number, reducing galvanometer current to $\frac{1}{10}$, $\frac{1}{100}$, $\frac{1}{1000}$ and $\frac{1}{10000}$ of
original value.

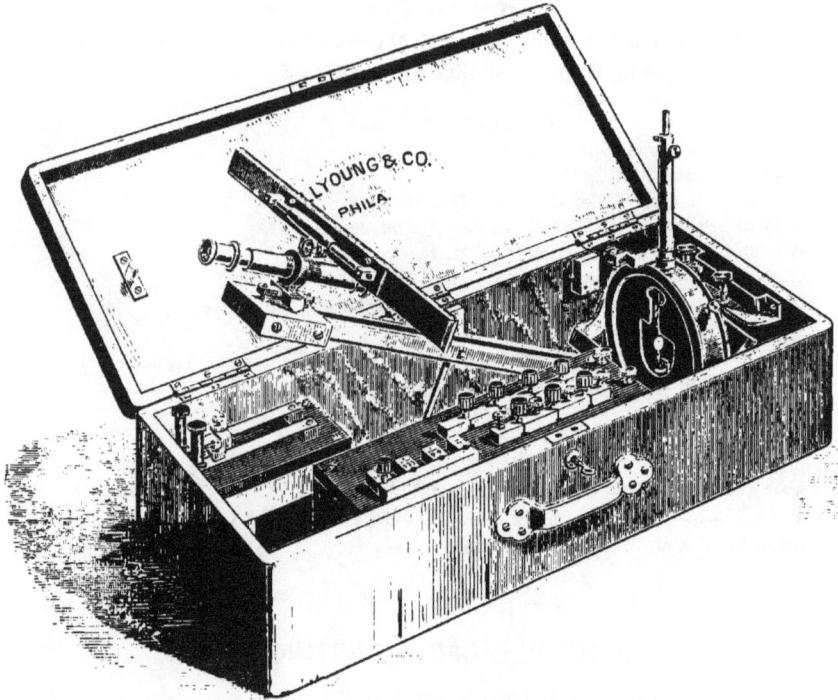

Fig. 85A.   W4570.

All the parts of the set possess the highest insulation, all metals and conductors being mounted upon hard rubber; the galvanometer, various coils, etc:, are dust and moisture proof.

With 100 volts battery, ten scale divisions are obtainable through 5000 megohms resistance.

The instrument as a whole folds up into a solid and highly polished oak case 22¾"x10"x7" deep and weighing about twenty pounds.   Case is equipped with lock and key and substantial carrying handle.

The entire time necessary to set up the instrument and have it ready for measurement with levelling, telescope adjustment and all effected, is not over one minute.

W4571. **Portable D'Arsonval Galvanometer**........ $100 00

Same as used in the Portable Cable Testing Set above described. Mounted in oak carrying case with lock and key, handle, etc.

W4572. **"Acme" Portable Cable Testing Set** $350 00

For insulation, capacity and simple resistance (conductivity) measurements on the street or in the laboratory.

This has but recently designed as is, we believe, one of the most complete and perfect outfits of its kind ever marketed.   The bulk and weight is a minimum weight is a minimum while the insulation and convenience of manipulation is a maximum.

The set consists of the following pieces:

1st Group.

W5003. 50 Cell Test Battery.

W4058. Condenser (five $\frac{1}{10}$ M. F. Sections).

W4340. Bridge and Resistance Set (without galvanometer).

W4364. Reversing and Discharge Key (Willyoung's improved).

Special. Reversing Switch for Battery.

Special. 10,0000 Ohms (two sections of 50 000).

W4201. Ayrton Universal Shunt.

All of the above in polished oak or mahogany case with handles, lock and key.

2d Group.

W4135. H. S. D'Arsonval Galvanometer.

Special small size, complete with scale and short focus telescope. This is in separate carrying case.

3d Group.

W4139. Dead beat tube for Galvanometer.

W4138. Ballistic   "   "   "

Six upper suspensions for Galvanometer.

Six lower   "   "   "

These items in separate case.

4th Group.

Tripod with extension legs.

Folding Stool.

In use the galvanometer fits upon the tripod head and connects to "1st Group" by flexibles. Manipulation of keys, plugs, etc., does not, therefore, cause any unsteadiness of the galvanometer.

The three cases of the groups above, together with the stool and tripod, all fit into a heavy leather carrying case provided with handles and carrying straps.

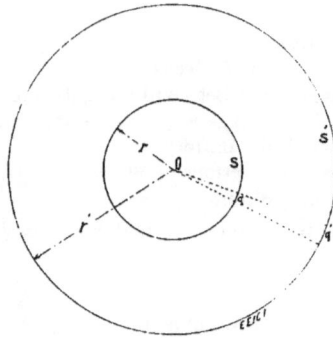

Fig. 86. DIAGRAM OF FUNDAMENTAL LAW OF PHOTOMETRY.

PHOTOMETERS AND PHOTOMETRY.

Photometry deals with the measurement of the light giving properties of light sources as also with that of the illumination thus produced.

Basic Law of Photometry.—Fig. 86 shows two concentric spheres of radii, r_1 and r_2. Their inner surfaces are uniformly white and a point source of light, L, is concentrated at the centre. If Q is the total light emitted by L; then a unit of surface of the inner sphere (area of surface $=4\pi r_1^2$) will receive a quantity of light

$$q_1 = \frac{Q}{4\pi r_1^2} \dots\dots(28)$$

Similarly a unit of surface of the outer sphere (area of surface $=4\pi r_2^2$) will receive a quantity of light

$$q_2 = \frac{Q}{4\pi r_2^2} \dots\dots(29)$$

Hence

$$\frac{q_1}{q_2} = \frac{r_2^2}{r_1^2} \dots\dots(30)$$

or *the quantities of light received upon parallel surfaces are inversely as the squares of their distances from the source.*

This law is commonly known as the **Inverse Square Law;** sometimes as the **Law of Distances.**

Total Intensity.—Consider equation (30) as applied to a sphere of unit radius; we shall have

$$q = \frac{Q}{4\pi} \dots\dots(31)$$

this is, evidently, a *constant* for any given light source. It is called the *Total Intensity,* and is generally designated by the letter l.

Intrinsic Intensity or Brilliancy.—Let the source be a small sphere, radius R, instead of a point; the light emitted by each element of surface, dS, will then be

$$dq = \frac{Q}{4\pi R^2}dS = \frac{I}{R^2}dS \dots \dots \dots \dots (32)$$

and, if dS = unity, $dq = \frac{I}{R^2}$. This represents the quantity of light emitted normally by unit area surface of the luminous source and is called the **Intrinsic Intensity** or **Brilliancy;** its symbol is i. Hence the *total quantity, Q, emitted by a luminous source is equal to the product of its brilliancy and its surface.*

The Unit.—Intensity of source must be measured in terms of the intensity of some specified source taken as a standard as e. g., a candle prepared in a definite manner, an incandescent lamp run at some specified voltage, etc.

Intensity of illumination must be measured in terms of the illumination produced by such an accepted light source when placed at some specified distance from the surface illuminated as e. g., a candle at one foot= the "candle-foot," etc.

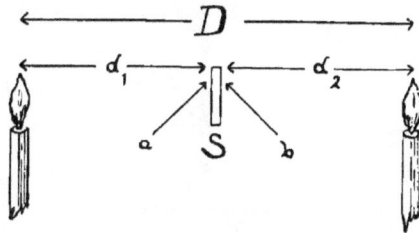

Fig. 87.

Measurement of Source Intensity.—Let L_1 and L_2 be two sources of light placed at a given distance, D, apart. S is a thin, opaque screen with both sides, a and b, uniformly white. Its plane is at right angles to the line joining L_1 and L_2. Move S to the right or left until both sides appear equally illuminated. Then if q_1 and q_2 are the quantities upon a and b due to L_1 and L_2 respectively, we have

$$q_1 = \frac{I_1}{d_1^2} \text{ and } q_2 = \frac{I_2}{d_2^2} \text{ or } \frac{I_1}{I_2} = \frac{d_1^2}{d_2^2} \dots \dots \dots (33)$$

And if I_2 be a standard source, K, we write

$$I_1 = K\frac{d_1^2}{d_2^2} \dots \dots \dots (34)$$

The Photometer.—This is a general term descriptive of the apparatus employed in making Photometric Measurements. While a great many basic schemes and consequent arrangements of apparatus are possible, nevertheless the Engineer and Electrician need only concern himself with those based on the plan just described (fig. 87) as thus far experience has proven them superior to all others for practical work.

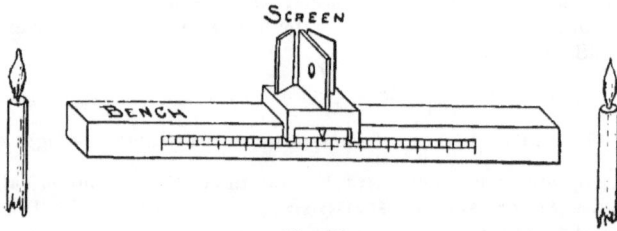

Fig 88.

The essential parts of the Photometer (see fig. 88) are the **Bench** (or **Track**), the **Scale**, the **Screen**, and the **Standard Light.** The **Bench** is the structure upon which the Screen is moved when getting a "balance." It must be accurately straight and placed level. The "Light Axis" (line joining the known and unknown source) must be placed so as to be accurately parallel to the bench.

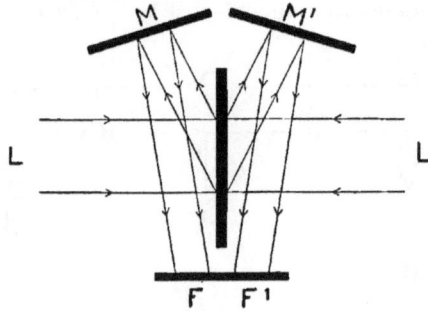

Fig. 89.

The **Scale** is mounted upon the **Bench.** It is uniformly divided into 100 or 1000 equal parts. Or it may be, and this is much to be preferred, an **Inverse Square Scale.** Such a scale is determined by using equation (34), calling K unity, and, assigning various values to I, determining the corresponding values of d_1 and d_2 into which the total bench length must be divided. These points are then marked off directly in values of *Intensity.* In use, therefore, balance is found and the reading multiplied by whatever may be the intensity of the standard [K of (34)].

The **Screen** (S of fig. 87) is a term used to cover the device upon which the two sources cast their light. It is placed upon a mounting adapted to move upon the bench; this mounting is called the **Carriage.**

The **Bunsen Screen** or **Sight Box** is the most commonly employed form of screen. It is shown in diagram in fig. 89. G is the screen proper—a sheet of thick paper with a star, circle, or other figure of transparent paper placed at its centre. This spot is usually produced by oil or paraffin and is, hence, called a *"grease spot."* Viewed from either side this spot shows dark against a light ground for its own light and light against a dark ground for the far light. The two mirrors, M and M¹, are placed so as to throw these two fields forward to the eye simultaneously. When they appear the same, the screen being moved to and fro until they do so appear, the screen is in "balance" and the illuminations due to the two lights are considered equal.

The Lummer-Brodhun Photometer Screen was devised by Drs. Lummer and Brodhun of the Reichsanstalt, and used in the photometric researches of that institution. In this instrument but one eye is used, thus eliminating any error due to the varying sensibility of the two eyes, and each of the sources of light being compared illuminates its own field, these two fields being presented to the eye as a disc and circle respectively, the latter surrounding the former, and having a sharp line of separation from it.

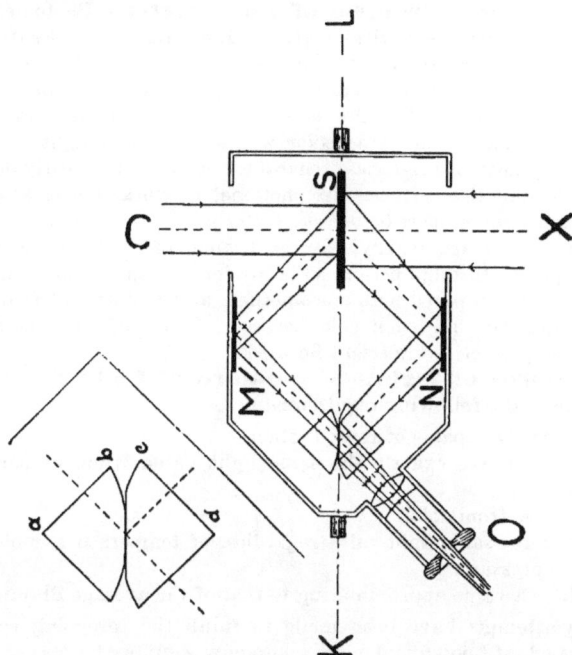

Fig. 90.

Fig. 90 shows the arrangement in diagram. C and X are, respectively the standard and the lamp being measured. S is an opaque white screen of plaster of Paris receiving lights on its two sides from C and X, from which it is diffused to the two mirrors M and N. P is composed of two right angled prisms whose hypothenusal faces are spherical surfaces cut off by a plane, more of one being cut than of the other; when in contact, therefore there is a disc of light from one source surrounded by a circle of light from the other sauce. Light from X normal to cd passes through both prisms where in contact, but suffers total reflection into O when it strikes the outer plane circle. Light from C, on the other hand, passes through both into O, where in contact, but is reflected back and out, (being absorbed by blackened surfaces) where it strikes the outer circle. An eye at O sees, therefore; a disc of light due to C, surrounded by a circle of light due to X.

This form of comparison apparatus is about four or five times as sensitive as any other form, and much greater accuracy can be obtained in a much shorter time. The average mean error of setting does not exceed 0.5 per cent. In Prof. Patterson's translation of Palaz, "Industrial Photometry" it is remarked that "the conclusion may be drawn that the optical screen of Lummer and Brodhun is at present that which presents the greatest advantages."

The General Problem of Photometry.—Photometry is, of course, not a strictly electrical matter. The value of all kinds of light sources, whether electrical or otherwise, are of interest and have often to be determined In the making of exact photometric measurements one of the greatest difficulties is that due to the light standard used not being of exactly the same *color* as the unknown light. Ability to determine equality of brightness regardless of color is greatly developed, however, by practice. Hence, in photometric work *one must expect to have to acquire experience before successful work can be done.*

Spectro-Photometry.—In this branch of photometric work both lights are passed through prisms so as to form a spectrum. The various colors are then compared with one another, and a curve plotted showing the light intensity for each color region. This kind of photometry is, however, only practiced in scientific work.

Standards of Light.--A **Standard of Light** should, theoretically, satisfy the following requirements:

(a) Be easily capable of reproduction.

(b) Always give exactly the same light value when prepared in the same way.

(c) Be free from color.

(d) Give the same light value regardless of temperature, moisture and barometric pressure.

(e) Have a value approximating to that of the average illuminant.

Many attempts have been made to fulfill the foregoing conditions. Some of the best known and most commonly employed types of standard lights are discussed below.

The Standard Candle.—This was the first unit to come into general use and is still rather popularly considered to be the *only ultimate* standard. Intensities have, therefore, been expressed in "Candle Power" —written C. P.

The Standard Candle, according to the English specification, must be of spermaceti consuming 120 grains of material (wick and all) per hour; wick must be of three strands, each with 18 to 21 threads. The candle must be 10″ long, 0.9″ diam. at base and 0.8″ diam. at top, and weigh 6 to the pound. When burned under normal atmospheric conditions and a flame height of 1.8″ the light emitted is defined as of unit intensity= 1 C. P.

Later suggestions (E. G. Love, *Am. Gas Lighting Journal.* March 5, '94, p. 326) would further fix the melting point of the spermaceti and a method of chemically cleansing the wick.

The German Candle is more minutely specified but is approximately the same as the English candle. It is made under the supervision of the " Verein für Gasbeleuchtung " and sent out with their certificate. The material is paraffin rather than spermaceti and the standard flame height is 2″.

Experts are now agreed that the *candle is not a reliable light standard.* This is because of the large variation of intensity for small differences in the quality of the spermaceti or paraffin, for greater or less tightness of twist in the wick, for small differences in the height of the flame, $\frac{1}{16}″$ making from 3 to 5% difference in intensity e. g., and for small variations in the amount of moisture. The sensitiveness of the flame to air currents, the quick and continual change in brightness due to the continuous untwisting of the wick, etc., also produe error.

When the candle is used two observers are needed; one to take the " balance " observations, and the other to observe the flame height at the moment of "balance" as also to keep this flame height as constant as possible. The candles should be burnt in a " candle balance " so that the *rate of burning* may also be known. Within small limits the C. P. may be taken as directly proportional (1) to the flame height and (2) to the amount of material consumed.

The Methven Screen.—This is an orthodox Argand Gas Burner with flame 3″ high and straight glass chimney 2″ in diameter. The light from this flame is cut off by a screen, in which is a slot 1″×0.233″. Viewed normal to the slot this flame is *supposed* to have 2 C. P. irrespective of the gas pressure or its composition provided its illuminating power lie between 5 and 10 candles. A "Carburetter" has been devised by means of which gas of any C. P. is claimed to give the same result.

This screen has the merit of great convenience and easy regulation. A correction must be applied for internal reflection from the chimney wall; this correction varies with the distance.

The Pentane Lamp.—This is a special form of lamp burning Pentane (C_5H_{12}), a hydro-carbon. The flame burns inside a metal chimney with an appropriate opening in the centre and a gauge for accurately setting the flame height. The pentane volatilizes at a temperature so low that the wick does not char and thus alter the rate of capillary supply of the combustible.

The principal difficulty with this lamp is that of obtaining pure pentane commercially. The lamp has been highly commended by an investigating committee of the British Association. The flame is a very pure white.

The "Amyl-Acetate" or "Hefner" Unit.—This lamp has been adopted officially by the National Electric Light Association (Niagara Meeting, '97) and, provisionally, by the Am. Inst. El. Engrs. (Proc. A. I.

E. E., '97, p. 90). The latter body says, "The Hefner Alteneck Amyl Acetate Lamp * * * should be temporarily adopted as a concrete standard of luminous intensity. or candle power."

The lamp (seen at the right in fig. 91) is a simple cylindrical reservoir of metal carrying a vertical wick tube of german silver. This wick tube must have a certain definite height, diameter and wall thickness. The flame should burn so that its peak is *exactly* 40 ohms above the top of wick tube; this height is determined by an optical "flame gauge" attached to the lamp. When burned at this height and using the proper quality of Amyl Acetate the relation below is said to hold

1 English candle = 1.136 Hefner Units.

The Amyl Acetate should satisfy the following: (1) Sp. Gravity @ 15°c = 0.872 to 0.876; (2) distilled in glass not less than 90% shall pass over for temperatures between 137°C and 143°C; (3) reaction neutral and blue litmus not sensibly reddened; (4) with equal amount of benzine must not become turbid; (5) c. c. of Amyl Acetate, 10 c. c. of 90% alcohol. and 10 c. c. of water must give a clear solution; (6) a drop evaporated on white blotter must leave no spot. Experience has shown, however, that pure Amyl Acetate is readily obtainable from reliable chemical houses.

Various Corrections.—The intensity of the Amyl Acetate Unit increases 2.5% for each mm increase of flame height and decrease 3% for each decrease of flame height. There is also an humidity correction expressed by the equation,

$$I = 1.049 - 0.0055h. \quad \dots\dots\dots\dots\dots\dots\dots(35)$$

when h is the humidity in liters of moisture to the cu. meter of dry air and I is the corresponding illuminating power.

For increase or decrease of barometric pressure we must add or subtract ¼% for each inch of pressure from the normal of 30".

To sum up, the **Amyl Acetate Lamp** *is unquestionably the most reliable and convenient light standard* thus far suggested. For this reason, as also on account of its small numerical value. it is best used as a primary standard only to be occasionally compared with the greatest accuracy with some other, larger valued and proper colored light, which shall be the **Secondary** or

The Working Standard.—The preceding standards are Primary or Ultimate Standards. In actual practice a **Working Standard,** itself occasionally carefully compared with the Standard. should be used. In this way we may choose a form whose color is the same as that of our unknown light and whose intensity value also is of approximately the same value. In Incandescent Lamp Photometry an incandescent lamp of the same nominal C. P. as the unknown, run at slightly below C. P., and with its C. P. in one direction exactly determined makes the most satisfactory Working Standard.

The Photometry of Incandescent Lamps.—The C. P. of an incandescent lamp varies with the voltage at which it is run. A voltmeter must be constantly employed, therefore, during such measurements, first to see that the voltage does not vary during the determination and, further, in order that the exact voltage corresponding to the measured

C. P. may be known. A rheostat capable of finely varying the applied voltage should also be employed.

The C. P. of an incandescent lamp also varies with the direction in which it is taken. Here the Universal Socket (W4602) may be employed; this allows measurements in a large number of directions to be made and the corresponding values may be later laid off in appropriate curves. A more convenient arrangement and one now almost universally in use among the lamp manufacturers and large users is the Rotator (W4594). This allows the lamp to be continuously spun on a vertical axis during the measurement so that the latter gives, immediately, the **Mean Horizontal C. P.**, or. if the rotator, W4596, is used the **Mean C. P.** for all zones varying by 5° from the horizontal may also be had.

The Photometry of Arc Lamps.—The C. P. of an arc lamp depends upon quality of carbons, voltage applied, air pressure, etc. Hence, photometric measurements are not satisfactory except for purposes of investigation or in connection with the development of improved types of lamps.

In Arc Photometry, to avoid removing the arc lamp too far, its light must be cut down in a known ratio. This may be done either by concave lenses (see *Ayrton*) or by a disk with adjustable open sector revolved so as to alternately intercept and transmit the beam.

To get the arc C. P. in axis other than horizontal the lamp may be arranged to describe a circle (at the extremity of a proper arm) in a plane perpendicular to the bench, the center of such circle being the axis of lights. An inclined mirror at the center and moving with the arm keeps the light rays always down the track upon the screen.

The Photometer Room.—A special room for photometric work is desirable but not necessary. Its walls should be, preferably, dead black and it should be well ventilated without drafts. Lights easily turned off and on should be provided, but should be so arranged as not to throw their rays directly into the eyes.

Choice of a Photometer.--Every photometer no matter how simple or cheap, must satisfy the following:

(1) Distance between lights must be accurately determinable.

(2) Adjustments to bring the screen and both lights into the same horizontal line, itself parallel to the scale, must be provided.

(3) The index and scale must give the true position of the central plane of the screen proper with reference to the two lights being compared.

(4) The "Track" or "Bench" must be straight; in this way the plane of the screen will always remain strictly parallel to itself.

THE WILLYOUNG PHOTOMETERS.

These are made in a variety of types as listed below and to satisfy all average demands from those of the occasional user to those of the large manufacturer or central station. Photometers intended especially for investigation and the most exact work are also provided.

All of the Willyoung Photometers have been most carefully worked out in every detail and will be found strictly as represented. That they represent real merit is evidenced by the fact that among purchasers may be mentioned: The Case School of Applied Science, The Edison Illuminating Co. of New York, The Narragansett Electric Light Co., The Zanesville Electric Light Co., The University of Vermont, Worcester Polytechnic Institute, Harvard University, Adelbert College, Colorado School of Mines, Ohio State University, Lewis Institute, Yale University, University of Virginia, Clarkson School of Technology, Washington Agricultural College, etc.

The "Reichsanstalt" Photometer is the form devised, adopted and recommended by the Imperial Reichsanstalt in Berlin. It is essentially a laboratory instrument, intended for the most exact work, and most suitable for the use of those making a scientific study of illuminants. The list of accessories which may be fitted to this instrument is large, and each accessory is provided with all possible adjustments.

The Willyoung No. 1 Station Photometer has been designed especially for large users, central station, manufacturers, engineering schools, etc., where a first class, all around instrument, capable of high accuracy and having all the conveniences for a large quantity of work is desired.

Fig. 91. W4581 W4582. REICHSTANSTALT PRECISION PHOTOMETER.

The Willyoung No. 2 Station Photometer is a more simple and less expensive instrument than the No. 1, although comprising essentially the same elements. It is made with accuracy and may be relied upon to do excellent work with considerable rapidity. To those not wishing to go to the expense of the No. 1 instrument, and to others who only require a photometer to make occasional measurements, we confidently recommend this instrument.

W4581. **Reichsanstalt Precision Photometer.** $195.00

Two meters (80 inches) between lights. Comprising following parts:

(a) *Bench*, consisting of two cold-rolled and absolutely straight steel shafts brought accurately parallel to one another and in the same plane and supported upon a number of suitable castings, each of which is provided with leveling screws working in a socket to be screwed to the table or brackets upon which the instrument is to be mounted.

A metal scale is screwed to the front bar of the bench and is engine divided, the divisions being in white on black ground. The graduations are in mms. and are numbered in both directions.

(b) Amyl acetate Standard Lamp with Reichsanstalt certificate.)

(c) Lummer-Brodhun Comparison Screen.

(d) Platform for Amyl Acetate Lamp.

(e) Socket for Standard Incandescent Lamp.

(f) "   " Incandescent Lamp to be measured.

(g) Two End carriages.

(h) One observing carriage.

The end carriages (g) are for the "standard" and "unknown" illuminant respectively. They thus take (d), (e) and (f). These carriages have both a coarse (clamp) and fine (rack and pinion) adjustment by which the pieces they support may be raised or lowered so as to bring the two lights and the center of the observing screen all in the same axis and parallel to the plane of the bench. Guard clamps, preventing the carriages from being accidentally knocked from the track, are provided.

The observing carriage takes (c), the Lummer-Brodhun comparison screen—or any other form of screen which it may be desired to use. A slow-motion screw is provided for effecting fine adjustment along the track. For reading the scale there is a small shaded incandescent lamp, together with key and binding posts. Normally this little lamp is shunted out of circuit; by depressing the key the scale is temporarily illuminated at the index.

W4582. **Reichsanstalt Precision Photometer** $217 00

Three meters (120") between lights ; otherwise same as W4581.

W4584, **Willyoung No. 1 Station Photometer** $175 00

Two meters (80") between lights.

The bench proper is made in the same manner and to the same dimensions as that of the Reichsanstalt Precision Photometer' All the fittings and accessories used with the Reichsanstalt form may therefore be used with this form.

The scale is engine divided on brass, white on black background—and is in inverse squares—reading directly in candle power.

Two graceful, cast pillars support the bench at each end. On the one bench is the rotator (W4594), while the other carries the socket for the Standard Incandescent Lamp. Binding posts are provided for tapping off voltmeter connections from the two lamps, as also for inserting an ammeter in the circuit of the "unknown" lamp.

On the front of each pillar top is a finely adjustable sliding rheostat for varying the voltage of the two lamps.

The motor driving the rotator is mounted on the pillar as shown; it has a rheostat for varying its speed.

Both rotator and standard lamp holder have vertical adjustments for their lamps by means of which their centres of illumination may be brought into the axis of the observing screen.

The screen is an improved Bunsen screen with specially ground mirrors. It is reversible as a whole so as to eliminate difference errors of the two eyes, the two sides of the screen, mirrors, etc.; the screen proper ("grease spot") is also independently removable and reversible. The carriage by which it is supported has no vertical adjustment as in the Rotator and Standard holder. A small shaded lamp just over the index lights up the scale momentarily upon depressing the little key at its side.

The combination of screens cuts off all light except that which goes straight down the track to the observing screen.

To sum up the Willyoung No. 1 Station Photometer comprises the following parts:

Bench mounted upon two cast pillars.	Rheostat for Standard Lamp.
Engine divided candle power scale.	Carriage for Bunsen screen,
Rotator W4594.	Bunsen Screen (reversible).
Motor for driving Rotator.	Two Track Screens.
Holder for Standard Lamp.	Two End Screens.
Rheostat for Rotator Lamp.	Two Standard Incandescent Lamps.

W4586. **Willyoung No. 1 Station Photometer**........ $190 00
Three meters (120") between lights. Otherwise same as W4584.

W4588. **Willyoung No. 1 Station Photometer**........ $215 00
The same as W4586 except that the Lummer-Brodhun Photometer Screen is substituted for Bunsen Sight Box; also Rotator W4596 is substituted for W4594.

W4590. **Willyoung No. 1 Station Photometer**........ $230 00
Three meters (120") between lights. Otherwise same as W4588.

W4592. **Willyoung No. 2 Station Photometer,**
(without motor.) $67 50
Two meters (80") between lights. Bench or photometer bar is of cold rolled steel braced with iron and japanned; it carries a brass scale, engine divided, to read direct in C. P,; white divisions on black ground. Scale is set at an angle of 45° so as to make reading easy without stooping. The whole is supported by a vertically adjustable pillar.

The screen is an improved Bunsen with best mirrors and reversible for elimination of differences of eye sensitiveness, mirror inequalities, etc. It mounts upon a simple carriage, moving without friction upon the top of the bar.

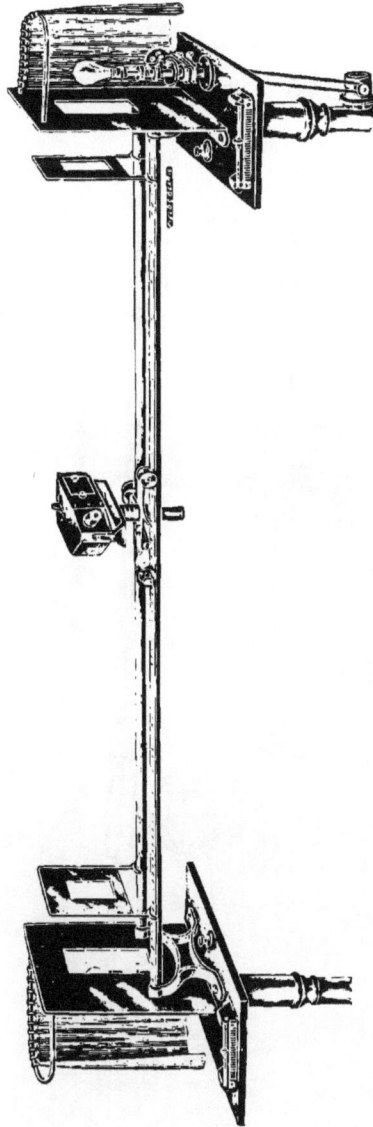

Fig. 22. No. 1 Station Photometer.

108

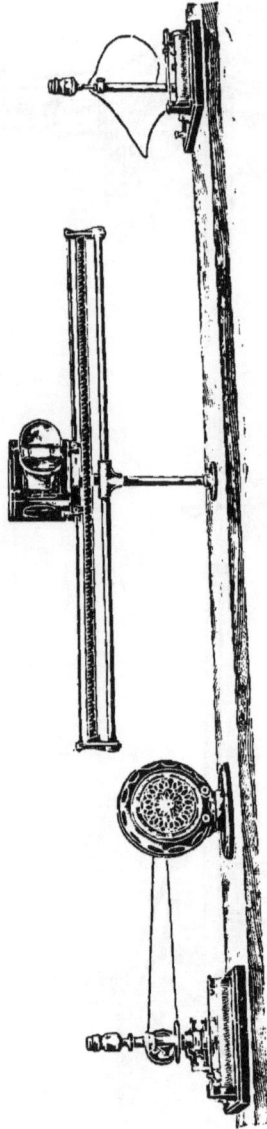

Fig. 93. No. 2 Station Photometer.

Rotator and Standard Lamp Socket are each mounted upon a small sub-base, each of which carries as well a sliding rheostat for bringing the voltage to the desired value.

In setting up the instrument the several parts may be mounted upon one table top or, if preferred, bar, rotator and standard lamp each upon a separate bracket or pedestal. (Table shown in Fig. is not furnished.)

In reversing the Bunsen screen (or any other screen of the Willyoung make) the screen turns upside down, and not as in most screens previously used, through 180° of horizontal arc, in the latter case requiring the photometer far enough from the wall for the observer to stand on both sides of it. The Willyoung Photometers may all be placed very close to the wall so as to economize space to the greatest possible degree. One "Standard" Lamp of 16 C. P. is supplied.

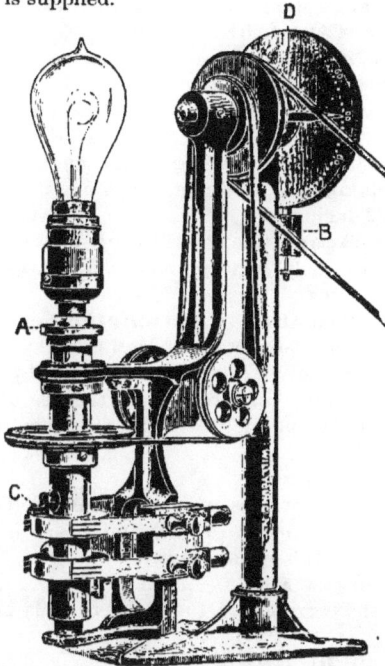

Fig. 94. W4596.

W4594. **Rotator** $30 00

For obtaining Mean Horizontal Candle Power; complete and ready for belting to driving motor. The lamp rotates upon a vertical axis driven by a small motor or other outside source of power through an intermediate friction pulley. A simple variable clutch allows this pulley to shift along its friction plate thus *varying the speed within wide limits and without stopping the rotator.*

The lamp socket is vertically adjustable so that any size of lamp may have its "center" brought into the "axis of lights."

W4596. **Universal Rotator**................ ......·........... .. $40 00

Same as W4594 except that axis of rotation may be set at any desired angle (by steps of 5°) with the vertical, thus allowing the mean C. P. to be obtained in any zone. A positive pin clutch B catches the frame at each desired angle. These angles are all variable and obtainable without stopping the rotation.

W4598. **Arc Light Mirror**............................... $80 00

For W4580 to W4592 inclusive for use in *arc light work*, with divided arc and circles.

W4600. **Dispersion Lens**................................ $30 00

For cutting down the intensity of arc lights. Three lenses of different foci with adjustable holder and screen and plane glass of same composition, for determining the absorption. For W4580 to W4592 inclusive.

W4601. **Rotating Adjustable Sector.**

For reducing the intensity of arc lights in a known ratio.

W4602. **Universal Lampholder**.......................... $90 00

For holding an incandescent lamp whose C. P. curve is being obtained. Rotation of the lamp about two perpendicular axes is provided for, there being divided circles for exact reading of the angles; in this way every possible face of the lamp can be directed against the comparison prism. For W4581 to W4592 inclusive.

W4604. **Lummer-Brodhun Photometer Screen.** $60 00

As described previously, fig. 90—ready to slip into the carriage of W4581 to W4592 inclusive.

W4606. **Lummer-Brodhun Photometer Screen**........$75 00

Same as 4604 but with two totally reflecting right angled prisms substituted for the usual reflecting mirrors as suggested by and made for Prof. B. F. Thomas of Ohio State University. By the use of these prisms the possibility of unequal absorption, always present when mirrors are used, is entirely done away with.

B5395. **Candle Balance** $45 00

Arranged to fit the carriages of Photometers W4581-W4590. Either one or two candles may be used. These candles are individually adjustable for height, as is the balance as a whole.

W4608. **Bunsen Sight Box**...................·............$25 00

Arranged to fit the carriages of the Reichsanstalt and the No. 1 Station Photometers.

W4608. **Bunsen Sight Box** $18 00

Ready to be attached to any style of carriage.

W4610. **Amyl Acetate Lamp**..............................$25 00

As previously described. Made in Germany and standardized for us at the Reichsanstalt, Berlin. It is supplied with the Reichsanstalt certificate of accuracy.

The Am. Inst. of El. Engs. have recommended that "the Hefner-Alteneck Amyl Acetate Lamp furnished with test certificates from the Physikalisch-Technische Reichsanstalt at Charlottenburg, Berlin, should be temporarily adopted as a standard of luminous intensity or candle power."

(See *La Lumiere Electrique, Vol. X, p. 501; Elek. Zeitschrift, Vol IV, 1883, Vol. III, p. 20, 1884.*)

W4612. **Chemically Pure Amyl-Acetate** $2 50
For use with the Amyl-Acetate Standard (W4610); per pint.

W4614. **Standard Sperm Candles** per lb. (6 candles).......$2 50
Best English manufacture, and as used in English practice, each. 50

W4616. **Standard Incandescent Lamps,** per pair.......... 2 00
Carefully standardized 16 candle power incandescent lamps for use as a secondary standard in electric light photometry.

W4618. **Methven Screen**...................................$38 00
With two slots. For ordinary illuminating or carburetted gas.

Unless specially called for all lamp sockets in any of our Photometers accessories will be of the Standard Edison type. If so ordered any other standard make of socket will be substituted without additional charge.

W4619. **Weber's Portable Photometer**.......... .... $175 00
This instrument is very easily manipulated and gives very satisfactory and accurate results. Large numbers are used throughout Europe especially. Three kinds of measurements may be made with the aid of this photometer, viz.:

FIRST.—Measurements of the intensity of sources of light, such as flames, etc., having the same color as the standard.

SECOND.—Measurements of diffused light having the same color as the standard.

THIRD.—Measurements of intensity, which light is of a *different* color from the standard.

The standard is a small benzine lamp and the screen a Lummer-Brodham.

The instrument is small and compact and does not weigh over 20 or 25 lbs. It may be easily carried from place to place and serves excellently for street measurements of arc lights.

A pamphlet describing this instrument in detail will be sent on application.

DIRECT READING (INVERSE SQUARE) SCALES.

The scales listed below have been computed with the greatest care. They are engine divided and are guaranteed absolutely correct. They read from $\frac{1}{50}$ to 50 *i. e.*, if one candle be the standard the bar will read from $\frac{1}{50}$ to 50 C. P. If the standard be a 16 C. P. lamp the bar will read from $\frac{1}{50}$ x16=.32 to 800 C. P., etc. Of brass.

W4620. **Two Meters Between Lights**....... ....$10 00
Graduations and numbers in white on dead black background.

W4622. **Three Meters Between Lights**................. 12 50
Graduated in same manner as W4620.

W4624. **Two Meters Between Lights**.................... 10 00
Graduations in white on black background.

W4626. **Three Meters Between Lights**............... .. 12 50
Graduated in same manner as W4624.

JUST OUT.

THE "AONE" PORTABLE PHOTOMETER.

For some time past there has been an increasing demand on the part of salesmen, small stations, engineers doing occasional testing, etc., for a small but well made Photometer, reasonable in price, compact and light, and capable of giving good commercial results on incandescent lamps. We have designed this instrument with a special view to this demand and believe that it properly fulfills these requirements. The whole instrument is ordinarily contained in a rectangular leather covered case measuring 20"x10"x8" and weighing about 25 pounds; handle and leather sling strap are provided. In use the lid is raised and thrown back; the two ends fall down; both ends slide into the bottom and may now be pulled out until motion is limited by the stops provided. On one end is the working standard which is a carefully tested 16 candle power lamp. The other end carries a small rotator upon which the lamp to be tested is placed. This Rotator is revolved by a small friction wheel attached to a hand lever. The screen is a modified Bunsen and the scale a direct reading scale. The hood of the Bunsen screen is turned upward slightly so that readings may be taken without stooping. Two curtains are arranged to throw around the lamps and thus keep the light from the eyes. The connections are so arranged that both the standard and the lamp to be tested are in parallel with the same circuit. If then the voltage of the unknown is the same as that of the known the result in candle power will not be seriously affected, if the working voltage is somewhat more or less than the designated voltage since the illumination may be assumed to vary for each lamp in the same proportion for a small change of E. M. F. If the unknown voltage is not the same as the standard the proper difference allowance may be made upon the small Rheostat forming a part of the apparatus The connections are easily changed at will, so that the two lamps may be run on separate circuits in the usual way, only, of course, a Voltmeter will be required to fix the voltage.

Two standard lamps are always furnished; one should be used as a working standard and the other should be carefully kept and only occasional reference made to it.

Adapters for Edison, Westinghouse and Thompson & Houston sockets are supplied.

W4630. **"Aone" Portable Photometer**................ $50 00

MAGNETOMETERS, DIP CIRCLES, AND OTHER INSTRUMENTS FOR MAGNETIC SURVEYS.

The demand for instruments of this character is so limited and the diversity of all admittedly excellent types so great that we have thought it best not to specifically *list* anything.

Apparatus of this kind is not regularly made by any American manufacturer. There are, however, several foreign concerns who have given special attention to this subject, working under the direction of eminent foreign authorities. We have taken pains to possess ourselves of all literature on this subject as it has appeared, and have complete catalogues. We shall be glad to go into the matter with those interested and to place our knowledge and catalogues at their disposal.

Our facilities for making such apparatus from special design are good and we shall be glad to give estimates at any time.

OTHER MAGNETIC APPARATUS.

W4632. **Hopkinson's Permeameter** $80 00

(See *Henderson*, p. 291.) A very massive wrought iron circuit has a cylindrical portion wound with two magnetizing coils. This cylindrical part is cut midway between ends so that one-half may be suddenly withdrawn; a small test coil fitting over it is thus released and jerked away by a suitable spring. Connected with a ballistic galvanometer the throw is, of course, proportional to the lines in the core. By making cores of the various irons we thus test permeability.

W4633. **Permeameter (S. P. Thompson's Form)**... ...$125 00

This is a work-shop instrument for determining permeabilities by measure of the tractive force required to detach two plane surfaces from one another when a magnetic field passes lines of magnetic force perpendicularly through the surfaces. The specimen to be tested slips through a hole in the top of a massive iron yoke and through a bobbin in which the magnetizing coil is wound. The lower end of the specimen is faced off and rests on a similarly faced part of the yoke. The traction required is measured by a spring balance. See *Jour. Soc. of Arts*, Sept. 12, 1890, also Ewing's *Magnetism in Iron and other Metals*, p. 248.

W4634. **Bismuth Spiral for the Measurement of Strong Magnetic Fields**, as suggested by Lenard.............$25 00

This instrument makes use of the well known fact that Bismuth changes its resistance in a magnetic field by an amount depending upon the strength of field. A one mm. Bismuth wire is wound non-inductively in a flat spiral about 30mms. in diameter and retained between two little flat discs of mica. The ends are soldered to copper leads passing down through a hard rubber handle to binding posts, from which leads may be taken to any Wheatstone's Bridge or other resistance measuring device. The change of resistance is about 5 per cent for a change of 1,000 lines of force in the magnetic field; this is given accurately by a calibration curve furnished with each inftrument.

This device is largely used in Europe among dynamo manufacturers for testing the fields of dynamos and motors, the exceeding thinness of the spiral allowing it to be readily slipped between any armature and pole piece even when the former is being driven at full speed. It will also prove useful in more refined laboratory work.

W4634A. **Snap Coil for Measuring Intensities of Mag-netic Fields,** as designed and made for Columbia College. $25 00

A coil 1 inch in diameter, of fine wire, is so mounted that it can be rapidly spun through 180° on release of a catch, earth inductor style. With a ballistic galvanometer this is very convenient for measuring the strength of dynamo and motor fields, etc.

"ARTIFICIAL LINES" FOR TELEGRAPH AND TELEPHONE INVESTIGATION.

We are prepared to supply Artificial Lines for experimental work in telegraphy or telephony and of any desired dimensions. These lines may be made either non-inductive with distributed capacity or inductive with distributed capacity. We are prepared to bid upon lines of any specification. We are also prepared to furnish Artificial Lines with distributed capacity and *self-induction* as proposed by Dr. Pupin (see *Proc. Am. Inst. El. Engrs.*).

W4635. **Artificial Line;** K. R. $= 10,000$. *Price on application*

This makes a good size line for telephone experimentation. The Line is separated in the middle so that each half may be used independently. Each half consists of 5 coils of 500 ohms each. At the terminals of each coil is joined one side of a Mica Condenser with a capacity of $\frac{1}{2}$ M. F. The K. R. value of each half line is, therefore, 2,500. The other side of each condenser has joined to its end a binding post. The whole is mounted on a polished case with a hard rubber top.

W4636. **Artificial Line, &c.** *Price on application.*

Same as preceding but with resistance and condenser joined up to blocks, so that individual resistances and individual condensers may or may not be used at will. Each condenser may or may not be short circuited by its own plug.

PENDULUM KEY SET.

For Complete Battery Tests, Condenser Tests, and Quick Electrical Measurements Generally.

This is a new instrument devised originally for use in our own Test Room. It is most useful in connection with any measurement which requires opening or closing, or both, or a number of circuits at certain intervals with relation to one another. In Condenser work, for example, where a Condenser must be discharged, short-circuited, etc., in succession and at definite intervals the work is much more accurate and entails a much less mental strain. *For complete Battery Tests, also,* where the Battery must be closed for a short time through resistance and a Condenser charged and discharged during this time, this apparatus is invaluable. Essentially there is a pendulum 18 inches long, pivoted, carrying two bobs, one above and one below the pivot, by means of which the time of swing may be widely varied. This pendulum is tipped with a steel paddle which sweeps over the arc of a brass circle the plane of which lies in the plane of swing. The keys, which may be either make—circuit or break-circuit, at will—are attached to this arc, and may be shifted along the arc as desired. By changing the positions upon this arc the relative time occupied by the paddle in getting from one to the other is changed. It is obvious that any number of circuits may be opened or closed, or both, in any desired order. The pendulum is released by a special key upon base, and is caught at the extreme of its swing by a special clutch. When turned over on the starting point the galvanometer is short-circuited and so remains until the pendulum is tripped.

The instrument is mounted upon a polished base and is well finished. Four "make" and four "break" circuit keys are furnished.

A photograph and working drawing will be sent on appplication.

W4637. **Pendulum Key Set.** *Price on application.*

W4638. **Electrical Contact Maker,** as supplied to the Navy Department for use in measuring acceleration in turrets of war vessels. The instrument is formed much like a tacheometer and consists of a geared Contact Maker enclosed in a case with main shaft, which may be thrust into the center of any shaft under test. Three sets of gears are supplied so that the electrical circuit is closed once, twice or five times for one revolution of the main shaft; or once for either one, two or five revolutions of the main shaft. The wedge of the Contact Maker is adjustable so as to allow for varying length of contact, and the brushes are also adjustable. A bracket is provided by means of which the instrument may attach permanently to the wall, and may be permanently connected as well to the shaft under test. *Price on application.*

FESSENDEN PORTABLE CONTACT MAKER.

W4639. **Fessenden Portable Contact Maker**...........$80 00

As designed by and made for Prof. R. A. Fessenden, of Western University, Allegheny, Pa., (See *Electrical World,* Dec. 5, 1896, p. 689), for quick and accurate tracing of alternating current curves. It is constructed much like the well-known portable tachometer with a spindle whose point is to be inserted in the center of the dynamo shaft. A pointer at the end of a vertical rod carries a vane dipping into a cup of oil and indicates the angle at which contact is made. By turning the handle by which the instrument is held this angle may be changed, the angle being read off upon the graduated degree scale. A condenser and Weston Voltmeter joined to the instrument in parallel give the corresponding E. M. F. values direct. With this instrument Prof. Fessenden states that "curves may be taken with a mean error for each ordinate of not over ½%, and in times greatly smaller than is possible with any other form of contact maker." The spindle is mounted on ball bearings and the instrument is otherwise of hard rubber and nickle-plated brass.

W4639A. **Improved Contact Maker** for taking alternate current curves with either ballistic galvanometer, electrometer, or electro-dynamometer................... .....$75 00

Insulation is very high so that the device may be used on high A. C. potentials with perfect safety. Has a universal coupler adapted to any sized shaft up to 2½" diameter. Of metal throughout nicely finished in japan and lacquer.

W4639B. **Automatic Contact Maker** $125 00

Similar to W4639A but with electro-magnetic rachet device to advance the angle; this angle is variable at will and is determined by the setting of a pin on the instrument. The advancing is effected by simple closure of a circuit from any point.

Further particulars and photograph on application.

W4639C. **Bedell-Ryan Jet Contact Maker** as devised and employed by Professors Bedell and Ryan. (See *Trans. Am. Inst. El. Engrs.*, Oct. 18, 1893.) *Price on application.*

Gives a perfectly constant and instantaneous contact, which is entirely free from the usual variations produced by exidation and wear of bearing surface.

We are the sole authorized makers of the Bedell-Ryan Contact Maker.

NORTHRUP OSCILLATING CURRENT GALVANOMETER.

FOR [HERTZIAN AND] RESONANCE [EXPERIMENTS AND FOR THE DETECTION OR COMPARISON OF HIGH FREQUENCY CURRENTS.

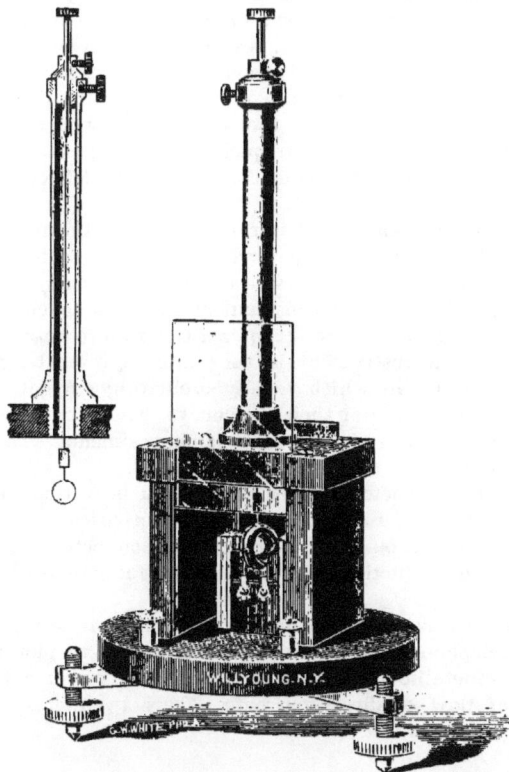

Fig. 95. W4640.

W4640. Northrup Oscillating Current Galvanometer.. $45 00

The working of this instrument depends upon the principle that when a metallic disc is suspended in a coil, the plane of the disc making with the plane of the coil an angle of about 45° the disc will tend to rotate, when alternating currents are sent through the coil, so as to increase this angle.

The instrument is constructed to be exceedingly sensitive, to have a minimum of self-induction, and practically no capacity. The disc is made of pure silver about $\frac{1}{64}''$ thick and 9 mm. in diameter. Three coils are furnished with each instrument. One coil has about 20 turns of No. 20, one about 40 turns of No. 42, and one of about one hundred turns of No. 36

B & S copper wire. Each coil is wound in two halves, so that the silver disc may be dropped down through the suspension tube and between the two halves of the coil. The inside diameter of the coils is about 1 mm. greater than the diameter of the disc. On either side of the hard rubber upright piece which supports the coils are the poles of a permanent magnet. The coils are set at an angle of 45° to the line joining the two poles and the silver disc hangs so that its plane is in this line.

The silver disc is fastened upon a light glass stem which carries a very small and thin mirror. This system is suspended upon an exceedingly fine quartz fibre. The complete period of swing of the system is about 12 secs., and the magnet quickly dampens the oscillations to zero. For small angles the deflections are proportional to the square of the current and to its frequency. Hence as long as the frequency remains constant two currents are to each other as the square roots of the respective deflections indicating them.

This instrument replaces and is far superior to the telephone in all cases where feeble rapidly varying currents are to be detected or compared.

The telephone fails to be of service when the frequency of the currents becomes very great; the present instrument responds to currents of any frequency including such as are set up in a Hertzian resonator. Since the self-induction of the instrument is very minute, it can be connected in series with any circuit in which rapidly oscillating currents are passing without appreciably changing their frequency. The instrument, therefore, serves in the performance of many Hertzian experiments. Thus it may be used to find the nodes along a wire.

Unlike the Electrometer its sensitiveness can be varied by non-inductive shunts, and it can be used with long non-inductive leads. It is especially adapted to experiments on electrical resonance; connected in a Wheatstone bridge made of non-inductive resistances it admirably serves for measuring liquid resistances.

It may be used like the telephone, but with currents so rapidly oscillating that the telephone would not respond to them, to plot the lines of current flow in metallic plates. These lines of flow would not correspond to the lines of flow of direct currents. Thus an interesting field for investigation presents itself.

Since the instrument responds only to alternating and pulsating currents, and not to direct, it can be used to determine if in any case a current is direct or pulsating. Thus the discharge of a storage cell through a low resistance is supposed to be more or less pulsating. The instrument would determine this point.

In short. we here offer an instrument which is nearly as sensitive for rapidly alternating currents as a Thompson galvanometer is for direct currents and many fields of investigation are thereby opened up. For special experiments with this instrument and its theory see *Electrical World*, Dec. 18, '97, etc.

WILLYOUNG HIGH POTENTIAL TRANSFORMER.

FOR PRODUCING RESONANCE PHENOMENA, HERT-ZIAN WAVES, VACUUM TUBE EXPERIMENTS, ETC.

W4645. **Willyoung High Potential Transformer** $75 00

Mounted in polished mohogany case, with double spark gap, etc. The transformer measures (outside) 6″ x 10″ x 11″ high and weighs 35 lbs.

Fig. 95A. W4645.

There is probably no single piece of apparatus which is of so extensive service in the performance of interesting electrical experiments as the high potential transformer. Every known experiment involving the use of oscillating currents, including Hertzian experiments on electric waves, Tesla's high frequency experiments, Pupin's resonance experiments, X-Ray and vacuum tube experiments, etc., can be best performed with the assistance of this piece of apparatus. While the transformer itself is somewhat costly and difficult to make, the subsidiary apparatus needed in the various experiments can in the main be "rigged up" on short notice and with slight expense; thus a physical laboratory provided with a high potential transformer is equipped for a very great variety of experiments.

With the single exception of vacuum tube and X-Ray experiments where a unidirectional discharge is required, the high potential transformer replaces and is in many ways superior to the induction coil. Its superiority rests upon the fact that the supply of energy from a transformer is continuous, while with an induction coil the supply of energy comes only at intervals which are great compared with the actual time of supply.

For a detailed description of a few of the experiments for which the transformer is useful, consult an article in the *Electrical World*, " On Induction with Currents of High Frequency at Long Distances," December 18th and 25th, by Dr. E. F. Northrup.

The transformer herein described is put up in a cherry box, highly polished It is provided with a double discharge gap between balls made of a non-arcing metal. This gap is enclosed in a hard fibre box so that the discharge is practically inaudible.

The object in making the discharge gap double is to make the oscillations which will accrue in a circuit in series with it more forcible than they would be were the gap single. However, in some Hertzian experiments it is desirable for obtaining the best oscillations that the partial arc which forms across the discharge gaps should be forcibly blown out. This requirement is provided for. The glass windows upon opposite sides of one of the gaps unscrew and in the place of one of them a tube to which a rubber hose can be fastened is inserted. The air blast can then be given by a foot bellows or furnished by means of running water.

In working with the transformer it is often desirable to obtain the high potential and still have a small current flow in the circuit of the secondary of the transformer. The result is obtained by inserting a small condenser in the circuit. The cover of the transformer box is provided with three binding posts between two of which such condensers may be inserted. Likewise resistances or inductance. The primary coils end in four binding posts by means of which they may be connected up in series or multiple, according to the potential of the alternating circuit employed. The current in the primary can be cut down as desired by external inductive resistances. The insulation of the transformer is very high so that a ¾" discharge can be drawn from its secondary without danger. It is so designed that the heating is insignificant and the current required very small. It must be understood, however, that if the secondary discharge is passing as an arc the secondary is practically short circuited, and therefore the primary will take an undue current. It is always well, therefore, in using the transformer to have a condenser in series with its secondary. This condenser should have a capacity as great or greater than the condenser, which may be used in the outside circuit in series with the discharge gap. In this case there will be scarcely any arc formed in the discharge gap and the length of the discharge between the balls will not be diminished.

CHLORIDE OF SILVER PORTABLE DRY CELL
TESTING BATTERIES.

These batteries are highly recommended for "testing" purposes. The cells are hermetically sealed, so that there is no spilling of acid. Any number of cells from 1 up to maximum capacity of battery can be used, by means of connecting cord provided. With exception of W5007-W5009 each battery is equipped with a current reverser. The E. M. F. is about 1 volt per cell.

W5000.	100 Cell Capacity; weight about 15½ lbs.						$100 00
W5001.	84	"	"	"	"	14 "	84 00
W5002.	60	"	"	"	"	10¼ "	60 00
W5003.	50	"	"	"	"	10 "	50 00
W5004.	40	"	"	"	"	9½ "	45 00
W5005.	32	"	"	"	"	8¾ "	38 50
W5006.	24	"	"	"	"	8 "	28 00
W5007.	16	"	" without current reverser				18 00
W5008.	12	"	"	"	"	"	13 50
W5009.	6	"	"	"	"	"	9 00

Cost of renewing and restoring cells when exhausted, per cell..... 35

Fig. 97. W5000, etc.

INDUCTION COILS AND X-RAY MACHINES.

Prior to the advent of the X-Ray an Induction Coil was, in general, a very inefficient and badly constructed piece of apparatus. Used, as it was, only occasionally for physical demonstration or spectroscopic work almost anything would answer and *did* answer. The demand for coils and upon coils was small and the incentive to improvement equally small.

With the X-Ray came the necessity for coils built to stand maximum strains for long periods without injury, to work with a minimum supply of energy, to have parts so made as not to get out of order, to have safety devices preventing accidental "burn-outs," etc., etc. In short, from a piece of interesting scientific apparatus the Induction coil had suddenly to become a *machine—an X-Ray Machine.*

Messrs. Willyoung & Co. were the first people in this country, if not, indeed, in any country, to look at the matter from the above point of view, and bent their energies at once to the accomplishment of this end. Their success was so great that for a while they had a practical control of the market. By degrees, however, and in accordance with the inevitable law of competition a part of this advantage had to be lost. It is a fact, however, that competitors were not able to approach our results with their own originally developed devices but only by adopting ours and the various radical improvements *effected by and original with* us are now used by all of our competitors who are making a thoroughly good product. Among these improvements original with Mr. Willyoung may be mentioned **The Adjustable Condenser, The Independent Multiple Vibrator, The Interlocking Switch, Improved Spark Points,** etc.

The Secondary.—But the most valuable and unique feature, perhaps, of the Willyoung Coils is the Secondary. This is wound in thin sections which are then assembled with very thin insulating septa between, after which the whole goes to a bath of special insulating composition, in which it is *cooked* for a number of hours; this drives out air and moisture. Finally a vacuum process removes the last remnant of air and cooling is effected so as to prevent contraction. The result is a hard mass of insulation containing the secondary, devoid of air and moisture, and which will not heat by the electrostatic bombardment of oscillating currents in the slightest degree. **The Willyoung Coils may thus be fully guaranteed.**

Choice of a Coil.—Good results are attainable in X-Ray work with any size coil from 4″ up. But the larger the better, inasmuch as the time required will shorten correspondingly. But we do *not* see any appreciable advantage in coils larger than 15″, although glad to make and sell them if called for. There is no tube made or suggested thus far which will begin to take care of the energy of a 15″ coil even.

We advise selection, therefore, as below:

Large Hospitals.............. ....12" at least, preferably 15"
Physicians and Surgeons in large practice .... 8"   "   "   12"
Physicians and Surgeons in small practice ..... 6"   "   "   8"

And, if you cannot afford better, then get a 4" as being better than none at all. As a matter of fact, however, the 4" is a very efficient coil and will give very good service either in X-Ray or scientific work.

In scientific work it is largely a matter of what is to be done and the scale on which it is to be done. *Any* size coil from 4" up will undoubtedly give good results in spectroscopic work, wireless telegraphy, Crookes phenomena, etc.

Battery or 110 Volts.—X-Ray Machines from 8" to 15" inclusive are supplied to run on storage battery or 110 volts direct as ordered. The 110 volt machines are to be preferred to battery machines where the necessary voltage is available, inasmuch as there is more energy available for the operation of the coil and the discharge, at any length, may be made as heavy as is desired while with battery, unless a large excess be used, the discharge is *not* usually the heaviest obtainable but merely heavy enough to make the rating of the coil an honest one.

*Fig. 98, W6000-W6004.

THE WILLYOUNG INDUCTION COILS.

These coils are nicely finished in polished cherry and equipped with our new "H. H." Interruptor, giving an extremely sudden break. The spark points also are of an improved type, moving freely and yet without slip or wobble and staying exactly where they are put.

W6000. **Willyoung Induction Coil**, 4" spark.............. $70 00
Requires two cells storage battery.
W6001. **Willyoung Induction Coil**, 6" spark.............. 82 50
Requires three cells storage battery.
W6002. **Willyoung Induction Coil**, 8" spark. 105 00
Requires four cells storage battery.
W6003. **Willyoung Induction Coil**, 10" spark............. 132 50
Requires five cells storage battery.
W6004. **Willyoung Induction Coil**, 12" spark............. 165 00
Requires seven cells storage battery.
W6005. **Adjustable Condenser**, added to any of the above
coils................................. .... 10 00

* Old cut. Photograph of latest model on application.

WILLYOUNG X-RAY MACHINES.

Fig. 99. 110-VOLT X-RAY MACHINE. W6014-W6017.

The coils listed below are really machines and are built for continuous and heavy service. They are equipped with every possible improvement, viz.: **Adjustable Condenser, Independent Multiple Vibrator, Interlocking Switch, Safety Fuse, Improved Parallel and Series Spark Gaps,** etc. The design has been thoroughly overhauled during the Summer just passed and a number of minor improvements effected. We have no hesitation in claiming that these machines cannot be surpassed in solidity, reliability, convenience of manipulation, and absolute perfection of working the world over.

W6010.	**Willyoung Battery X-Ray Machine,**			15" spark,	$252 50
W6011.	"	"	"	"	12" "	195 00
W6012.	"	"	"	"	10" "	167 50
W6013.	"	"	"	"	8" "	135 00
W6014.	**Willyoung 110-Volt X-Ray Machine,**			15" spark,	300 00
W6015.	"	"	"	"	12" "	240 00
W6016.	"	"	"	"	10" "	212 50
W6017.	"	"	"	"	8" "	180 00

The above machines are ready for operation *at once* upon attachment to battery or the 110-volt circuit as the case may be.

W4014. X-Ray Adjustable Condenser.................... $30 00

About 5 M. F.'s total of condenser is mounted in finely finished cherry case and equipped with one of our Improved Condenser Switches. *This condenser may be used with any coil whether or not of our make.*

W6020. Independent Multiple Vibrator................. $15 00

This is our latest pattern of Multiple Break and the same as used on the various machines W6010-W6017. It is mounted upon a finely finished cherry or mahogany base with marked binding posts and can be used with any coil of any make.

In ordering parties should give particulars of the coil with which it is to be used, as we can then wind the break accordingly and thus assure the best results.

W6021. Independent Multiple Vibrator and Adjust-
. **able Condenser** $40 00

Consists of W4014 in somewhat larger case and with W6020 mounted thereon as well.

W6022. Rotary Break.................................... $50 00

For Coils and X-Ray Machines of all sizes. A direct current motor operates a breakin mercury under oil. Speed of break is variable at will. Motor of any desired voltage.

W6023. Rotary Break.............................. .... $60 00

Same as W6022 but alternating instead of direct current motor.

Among other users of Willyoung X-Ray Machines and Induction Coils are the following:

PENNA. HOSPITAL, Philadelphia, Pa.
POLYCLINIC HOSPITAL. Philadelphia, Pa.
JEFFERSON COLLEGE HOSPITAL, Philadelphia, Pa.
EPISCOPAL HOSPITAL, Philadelphia, Pa.
JEWISH HOSPITAL, Philadelphia, Pa.
GERMAN HOSPITAL, Philadelphia, Pa.
GIRLS' HIGH SCHOOL, Philadelphia, Pa.
GIRLS' NORMAL SCHOOL, Philadelphia. Pa.
BOYS' HIGH SCHOOL, Philadelphia, Pa.
DR. CHRISTINE, Philadelphia, Pa.
JOHNS HOPKINS HOSPITAL, Baltimore, Md.
SOUTHERN HOMEOPATHIC MEDICAL COLLEGE, Baltimore, Md.
UNIVERSITY OF NEW YORK, New York, N. Y.
U..S. NAVY YARD, Brooklyn. N. Y.
EMERGENCY HOSPITAL, Washington, D. C.
MARINE HOSPITAL SERVICE, Washington, D. C.
CATHOLIC UNIVERSITY, Washington, D. C.
EASTERN HIGH SCHOOL, Washington, D. C.
McGILL UNIVERSITY, Montreal, Can.
MONTREAL GENERAL HOSPITAL, Montreal, Can.
ROYAL MILITARY COLLEGE, Kingston, Ont.
U. S. NAVAL ACADEMY, Annapolis, Md.
WILLIAMSPORT HOSPITAL, Williamsport, Pa.
LANCASTER GENERAL HOSPITAL, Lancaster, Pa.
BRYN MAWR COLLEGE, Bryn Mawr, Pa.
BUCKNELL UNIVERSITY, Lewisburg, Pa.
MICHIGAN UNIVERSITY (3), Ann Arbor, Mich.
DR. W. H. CRANE, Ypsilanti, Mich.
STATE NORMAL SCHOOL Ypsilanti, Mich.
DELAWARE COLLEGE, Newark, Del.

MEDICAL COLLEGE OF OHIO, Cincinnati, Ohio.
ADELBERT COLLEGE, Cleveland, Ohio.
PRESBYTERIAN HOSPITAL, Chicago, Ill.
ST. LUKE'S HOSPITAL, Chicago, Ill.
ARMOUR INSTITUTE, Chicago, Ill.
HOSPITAL, Alton, Ill.
WOFFORD COLLEGE, Spartansburg, S. C.
SOUTH CAROLINA COLLEGE, Columbia, S. C.
PORTER MILITARY ACADEMY, Charleston, S. C.
WAKE FORREST COLLEGE, Wake Forrest, N. C.
DAVIDSON COLLEGE, Davidson, N. C.
MISSOURI STATE UNIVERSITY, Columbia, Mo.
UNIVERSITY OF VERMONT, Burlington, Vt.
UNIVERSITY OF ALABAMA, Tuscaloosa, Ala.
NORTHERN INDIANA NORMAL SCHOOL, Valparaiso, Ind.
PURDUE UNIVERSITY, Lafayette, Ind.
UNIVERSITY OF NOTRE DAME, Notre Dame, Ind.
UNIVERSITY OF TENNESSEE, Knoxville, Tenn.
SOUTHWESTERN UNIVERSITY, Georgetown, Texas.
SMITH COLLEGE, Northampton, Mass.
IOWA STATE UNIVERSITY, Iowa City, Iowa.
KANSAS STATE AGRICULTURAL COLLEGE, Manhattan, Kan.
REINHART HOSPITAL, Ashland, Wis.

ELECTROLYTIC INTERRUPTERS.

Fig. 99A. W6025

These interrupters, of the Wehnelt or Caldwell type, have excited much interest since their recent discovery (Spring of 1899). Essentially they are most simple, consisting only of a small surface (3 or 4 sq. mms.) platinum anode and a large surface (200 or 300 sq. cms.) lead kathode immersed in dilute H_2SO_4. If joined in series with the primary of an induction coil and a sufficient E. M. F. exceedingly rapid (1,000 to 10,000 per minute or more) breaks take place at the platinum surface.

These breaks are *probably* due to the sudden formation of an envelope of non conducting gas about the platinum surface. Their frequency varies *directly* as the E. M. F. employed and *inversely* with the area of platinum.

In practice at least 40 volts must be used to obtain good results; on large coils requiring 4 or 5 amperes or more this means a *heating* loss of ¼ HP up. The interrupter is not, therefore, theoretically efficient.

When using these interrupters with a coil *no condenser is required.*

The secondary discharge is very thick and vigorous. Unless great care is used the "target" in the ordinary X-Ray tube will be burnt away. Special tubes are desirable.

On account of the large amount of secondary energy liberated exposures by use of the electrolytic interrupter may be very short.

For spectroscopic work the heavy discharge should be very useful.

The objections to the electrolytic interrupter are: 1, that it is fussy and mussy; 2, that it requires so high an E. M. F. However, if one is willing to be careful and clean he will find it an exceedingly valuable accessory and one which will well repay investigation and use.

W6024. **Wehnelt Interrupter,** for coils up to 15″ in spark length.............................. ... ...$25 00

Has three platinum anodes of different surface; these are removable at will and may be used singly or in multiple as desired. The whole is contained in a glass jar 4″ diam. by 5″ high, mounted upon a japanned iron base. The cover carrying the anodes fits tightly in the jar and prevents acid fumes being given off. For short runs only.

W6025. **Wehnelt Interrupter** $35 00

Same as W6024 but provided with a water or cooling jacket so as to permit of continuous running. May have running water from local supply system passed through it; or the jacket may simply be filled with a fresh supply from time to time.

W6026. **Silver Voltameter** of variable resistance............$25 00

Kelvin's form, but fitted with rack and pinion as suggested by Prof. Carhart. Two anode plates are hung on each side of a single cathode; the plates may be raised or lowered at will, so as to provide for slight variations of current. The plates are held in place by spring clips, and are readily removable.

STATIC MACHINES.

The machines here listed are made for us by a firm in whom we have confidence as being careful, up-to-date and reliable. Those of our customers desiring Static Machines, either for X-Ray or demonstration work, will make no mistake in choosing one of these machines.

We shall be pleased to quote special prices and give specifications to those desiring more elaborate and larger machines.

Fig. 99B. No. 1 Static.

In the machines listed below the woodwork is polished mahogany or oak as desired. A cover case of wood and glass is a part of every machine which is thus protected from dust and damp. A door in each end of this case gives access to the interior. All metal parts are supported on hard rubber pillars. A special driving gear permits of speeds from 1,500 to 2,000 per minute. If to be driven by a motor, a special pulley will be attached without extra charge.

No. 1 Static Machine..$35 00

One glass and one rubber plate. For school and experimental work. Gives 4″ to 5″ spark.

No. 2 Static Machine,
$52 00

Two glass and two rubber plates. Recommended for X-Ray and therapeutic work. Gives 8″ spark.

No. 2 (Special) Static Machine $67 50

Same as No. 2 but built more solidly throughout. Especially intended for long continuous runs with motor driving.

No. 3 Static Machine,
$100 00

Four glass and four rubber plates. For physicians and hospitals; quick X-Ray work, etc. Gives a very thick 8″ spark.

Pole Changer, attachable to any Static Machines, with aid of a screw driver, in five minutes............$8 00

Fig. 99C. No. 3. STATIC.

By use of this device the current through an X-Ray tube may be instantly reversed without changing connections.

Spark Gaps...................... ....per pair, $3 00

For Static Machines when used for X-Ray work. One is clamped to each discharge rod and each is adjustable. There is, therefore, a variable at will gap at *each* terminal of the X-Ray tube.

X-RAY TUBES.

As to what is the very best form of tube opinions differ widely. Perhaps, indeed probably, there *is* no *best* form; the whole matter is one of conditions, like everything else. The size of the coil, its general relation of parts, the generating source, the amount of current used, etc.—all these require certain differences in the tube if the *most* harmonious relations are to be secured. Still, there are certain landmarks determined by the designs of the coil and its size. The tubes listed here have been developed with particular reference to the design of our coils and may be depended upon to give excellent results.

These tubes are made either with or without adjustment for vacuum. The cuts figs. 100-101 both show this latter device, which consists of a side pocket containing a substance from which, by use of a match or spirit lamp, vapor may be driven off to reduce the vacuum where this shall have become too high. The device is efficient and is the next best thing to the "self-regulating" tube. The plain tubes (no adjustment for vacuum) are also good and may be kept in excellent condition as regards vacuum by merely heating the *bulb* of the tube with a spirit flame.

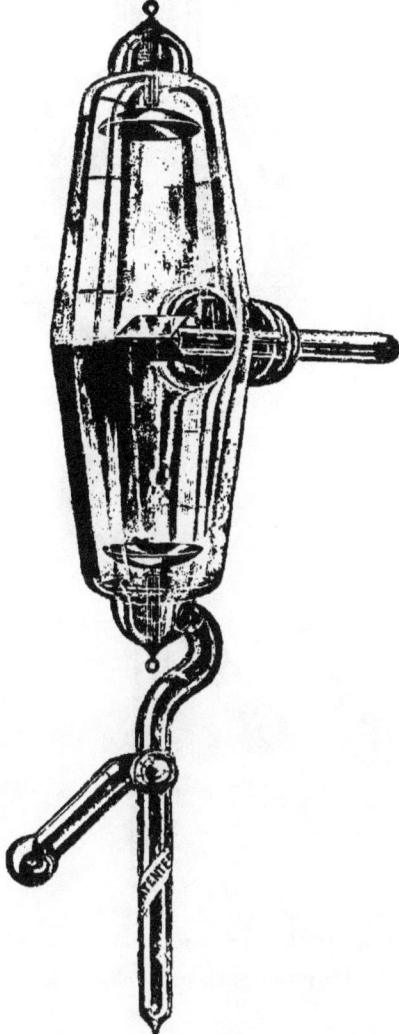

Fig. 100. W6630. THOMSON DOUBLE-FOCUS TUBE.

Fig. 101. SINGLE FOCUS TUBE WITH REGULATOR.

If without regulator the offshoot is omitted.

X-RAY TUBES WITH REGULATOR.

W6030.	**Thomson Universal Double-Focus Tube** (for Tesla coils only)					$12 00
W6031.	**Single-Focus Tube,** for	15″	coil			12 00
W6032.	"	"	"	" 10″ & 12″ "		10 00
W6033.	"	"	"	" 6″ & 8″ "		9 00
W6034.	"	"	"	" 4″ "		8 00

X-RAY TUBES WITHOUT REGULATOR.

W6038.	**Single-Focus Tube,** for	15″	coil			$11 00
W6039.	"	"	"	" 10″ & 12″ "		9 00
W6040.	"	"	"	' 6″ & 8″ "		8 00
W6041.	"	"	"	" 4″ "		6 00

QUEEN SELF-REGULATING TUBES.

The "penetration" of a tube seems to depend, principally, upon its vacuum or degree of exhaustion. With increase of vacuum we get better penetration, and this, though most desirable in hip-joint work e. g., simply wipes out everything when the tissue is thin as in hand and arm work. If the vacuum goes down even the flesh becomes dense and the Rays will go through nothing and the tube is worthless.

With the usual types of tubes this instability of vacuum often takes place after short use and constitutes one of the most serious defects met with.

The "Queen" Self-Regulating Tube was devised to provide against this difficulty by automatically maintaining the vacuum constant and also to enable any devised degree of penetration to be secured and this particular degree to be maintained. It is claimed that these objects have been very perfectly attained in this tube.

We have made arrangements with the manufacturers by which we can offer these tubes, adapted to Willyoung Coils, to our customers. For large coils, particularly, they are recommended.

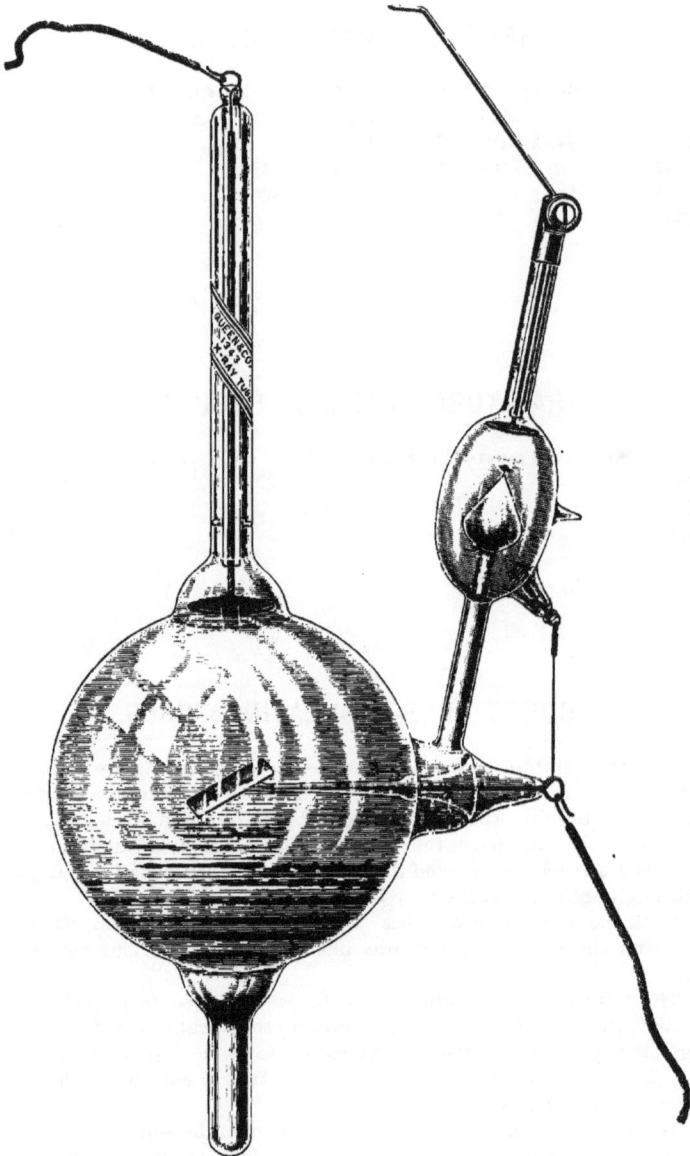

Fig. 102. W6042. SELF-REGULATING X-RAY TUBE.

DESCRIPTION OF SELF-REGULATING TUBE.

A reference to the illustration will make the operation of the tube clear. A small bulb containing a chemical which gives off vapor when heated and reabsorbs it when it cools, is directly connected to the main tube, and is surrounded by an auxiliary tube, which is exhausted to a low Crookes' vacuum. In the auxiliary tube the cathode is opposite to the above mentioned bulb, so that any discharge through it will heat the bulb by the bombardment of the cathode rays. This cathode is connected to an adjustable spark point, the end of which may be swung to any desired distance from the cathode of the main tube. The anode of the small tube is directly connected to the anode of the main tube. The coil is connected as usual to the main tube, which has been exhausted to a very high vacuum, and consequently has a high resistance. When it is put in operation the current takes the path of least resistance by the spark point and the auxiliary tube, and heats the chemical in the small bulb, thereby releasing the vapor which it contains in state of absorption and driving it into the main tube. This will continue until a sufficient amount of vapor has been driven into the main tube to permit the current to go through it, which will begin to take place when the vacuum has been reduced until the resistance of the main tube is brought down to that of the spark gap plus the small resistance of the auxiliary bulb. After this only an occasional spark will jump across the gap to counteract the tendency of the chemical as its bulb cools to re-absorb vapor and raise the resistance of the main tube. The tube is thus maintained at a constant vacuum while running. When the current is stopped the chemical cools off and re-absorbs vapor and the tube returns to its starting condition of high vacuum.

It will be evident from the above that the height of the vacuum at which the tube runs will depend on the resistance of the circuit through the auxiliary bulb; in other words, on the length of the spark gap. The tube may be set to run at high vacuum by placing the spark point at a considerable distance from the cathode terminal of the main tube, or to run low by placing it near. With the Self-Regulating X-Ray Tube the vacuum may be made so high that it is impossible to force any current through it, or so low that the current will go through the tube rather than jump over a parallel spark gap of one-half inch, a range of vacuum that includes all possible X-Ray work.

The cathode is of pure hard hammered aluminum, accurately ground and polished to the proper curve to concentrate the cathode rays on a small surface of the platinum. This gives a small source of X-Rays and insures radiographs of sharp definition and clearness. The cathode terminals of both the main tube and regulator are extended and protected by double thickness of glass, making puncturing almost an impossibility. The anode is of heavy platinum foil $1\frac{1}{8}$ inch x $\frac{7}{8}$ inch. Its thickness and size make it able to stand considerable energy without being melted through, and it is set at an angle of about 60 degrees to the path of the cathode rays, which increases the accuracy of definition and efficiency without materially diminishing the field of view. The lower part of the main bulb has an extension of strong glass, remote from the leading in wires, for clamping the tube. The bulb containing the regulating chemical is made conical, the point towards the cathode in order that there shall be an even distribution of the heat generated. The point of the bulb is protected by a small platinum shield.

PRICE LIST.

W6042. **Self-Regulating X-Ray Tube,** large size, with extra heavy platinum for use with coils giving very heavy discharge. $18 00

W6043. **Self-Regulating X-Ray Tube,** standard size for all ordinary work.................... .... 15 00

W6044. **Self-Regulating Double Focus Tube** for Tesla or Thompson coils 18 00

This tube has the two cathodes at right angles to each other and the two bundles of cathode rays impinge on the same side of a heavy platinum anti-cathode, thus giving better definition than the usual form in which the X-Rays originate on both sides of a platinum wedge.

X-RAY STANDS.

We make two styles of stands as below. The smaller is of wood and may be used upon any table or stand. It has every possible adjustment and will be found adequate for all ordinary demonstrative work, hand and arm examination, etc.

The larger stand has been designed especially for the needs of the surgeon and practitioner. It is very solid and mounted upon a heavy cast base. The structure is of steel bicycle tubing. The adjustments are such that the tube may be placed in *any desired position* within a cylinder 6 feet high and 6 feet in diameter. In surgical work this means under or over, behind or in front of a patient and at any desired angle. The holder for the tube is cork lined and the joints are universal so that the tube need merely be twisted to its proper position and one screw tightened to set the whole finally. The connecting wires are carried by our special "slip clamps" at the extremities of insulating rods; in case of any strain upon the wires the "slip clamps" automatically free the wires, thus avoiding any possible breaking out of the tube terminals.

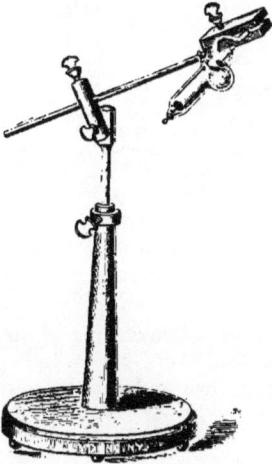

Fig. 103. WOOD STAND.　　　Fig. 104. UNIVERSAL STAND.

W6050. **Table X-Ray Stand**............ .... §2 50

　　Of wood as described.

W6052. **Universal X-Ray Stand**.......... .... 15 00

　　Of metal throughout; universal joints and insulating carriers for connecting wires as described.

FLUOROSCOPES AND FLUORESCENT SCREENS.

Fig. 105.

We are prepared to furnish both tungstate of calcium and platinum - barium cyanide screens in any size desired and mounted either with the hood or simply as plain screens. Our hood screens or **Fluoroscopes** proper as shown in the cut are now made so that the screen itself may be removed from the hood at will and used separately.

We recommend the platinum screen as well worth the additional charge necessitated by the cost of its valuable salt. The definition is much more brilliant and there is not the haziness of outline always found in calcium screens and due to the *phosphorescent* quality of the latter. There is, also, no deterioration in the platinum screen, whereas, in the calcium screen, the responsiveness of the same is practically lost in the course of a year or two.

PRICE LIST.

	With B. P. C. Screens.	With Tungstate of Calcium Screens.
3 x 4 inch	$ 6 50	$ 5 00
5 x 5 "	10 00	7 50
5 x 7 "	12 50	9 00
6 x 8 "	16 00	11 00
7 x 9 "	20 00	13 50
8 x 10 "	24 00	16 00

Unmounted Barium-Platino-Cyanide Screens, 25 cts. per sq. in.
Unmounted Tungstate of Calcium Screens, 15 cts. per sq. in.

We are prepared to make screens of any size and shape desired, and estimates will be cheerfully furnished.

DRY PLATES, ETC., FOR X-RAY WORK.

Almost any plate will *do* for X Ray pictures, but we consider the Carbutt plates much superior to any other brand. Mr. Carbutt has studied the problem of producing the *best* plate for this purpose, and we believe he has succeeded. The plate is extra thick in film and gives a good solid image, and does it quickly. Each plate is enclosed in a light tight envelope and is at once ready for use. No plate holder is required and no "dark room" until the exposure is over and development begins.

In ordering X-Ray plates merely designate the size and quantity, thus: "Send 1 doz. 10 x 12," etc.

Size.	Per doz.	Size.	Per doz.
4 x 5 in..................	$0 80	10 x 12 in..............	. $4 50
4¾ x 6½ "	1 20	11 x 14 ".................	6 00
5 x 7 ".......	1 30	14 x 17 "	11 65
5 x 8 "	1 50	16 x 20 "	... 15 90
6½ x 8½ "....	2 00	17 x 20 "	16 60
7 x 10 ".	2 50	18 x 22 "	21 00
8 x 10 "	2 90	20 x 24 ".......	23 40

Carbutt Metol-Hydro Powder, small size..... $0 25

For developing X-Ray plates.

Carbutt Metol-Hydro Powder, large size 50

This developer is simple, clean and handy; it does not deteriorate while awaiting use. It is put up in hermetically sealed glass bottles.

—

STORAGE BATTERIES FOR X-RAY WORK.

Any of the Willyoung Coils can be operated from primary batteries if so ordered. We earnestly advise, however, the use of storage batteries in preference as requiring practically no attention except occasional charging and as occupying less room.

The "Chloride" accumulator which we regularly supply is generally considered one of the best storage batteries made. They will last almost indefinitely with ordinary care in usage. They are portable, give off no fumes, and cannot easily be injured.

If purchased of the preferred sizes below a storage battery will give proper service for the ordinary run of work of the average hospital for a month or six weeks without recharging. When this is necessary the battery may either be charged in the office (provided it is wired with direct current) or it may be sent to the nearest direct current station. *The storage battery cannot, of course, be charged from an alternating circuit.*

The smaller batteries listed will do the work as well as the preferred sizes, but will not, owing to their smaller "ampere hour" or "total current" capacity, do the work for as long a time.

The batteries are all mounted in portable oak carrying cases with handle: *there are no glass jars to break.*

PORTABLE STORAGE BATTERIES.

PREFERRED SIZES.

Each individual cell in this group of batteries has a total capacity of 8 hours at 7½ amperes of current and a correspondingly *longer* time at a proportionately *less* current.

W6060.	**Portable Chloride Battery**, for 4″ coil (2 cells)...	$22 00
W6061.	" " " " 6″ " (3 cells)...	30 00
W6062.	" " " " 8″ " (4 cells)...	40 00
W6063.	" " " " 12″ " (5 cells)...	50 00
W6064.	" " " " 15″ " (7 cells)...	70 00

NON-PREFERRED SIZES.

Here each individual cell has a total capacity of about two-thirds that of the preferred sizes. For occasional use and for physicians and surgeons in small practice this size will be found adequate.

W6068.	**Portable Chloride Battery**, for 4″ coil (2 cells)...	$18 00
W6069.	" " " " 6″ " (3 cells)...	26 00
W6070.	" " " " 8″ " (4 cells)...	32 00
W6071.	" " " " 12″ " (5 cells)...	38 00
W6072.	" " " " 15″ " (7 cells)...	50 00

Unless otherwise ordered, portable batteries will be shipped filled with acid and charged ready for service. While the greatest care is used in packing, it is almost impossible to avoid damage to cells of this type when shipped by freight; for short distances we therefore recommend shipment by express, where packages receive more care in handling. Where this method is too expensive we advise shipment without acid, which may be ordered to be shipped in separate vessel, for which a small extra charge is made.

WIRELESS TELEGRAPHY.

Wireless Telegraphy is still in its infancy. But the possibilities indicated by the small amount of work which *has* been done have excited special interest and promises well for a still more rapid and practical development of the subject. We have been employed for over a year past in making much of the experimental apparatus used in this country on this class of work; practically all of that used by the Ordnance Department and the Signal Corps *e. g.* has come from our shops.

A large factor in the successful operation of the apparatus being a suitable induction coil, and our coils holding a reputation second to none, has brought us many inquiries.

We have no "system" of wireless telegraphy and doubt whether there is any "system." But we are prepared to furnish coils, coherers, relays and oscillators and all other portions of outfits which may be required.

We have made arrangements with Mr. W. J. Clark by means of which we are enabled to offer a simple form of demonstration apparatus which is sufficient for schools and students to demonstrate the fundamental sequence of the involved phenomena. This outfit is described and listed below:

W6073. **Small " Wireless " Outfit**...........................$40 00

Consists of transmitter, induction coil complete with oscillator, battery and key. Also receiver, consisting of high resistance relay, decoherer, coherer, Buzzer and battery, suitable for transmissions of 25 to 100 ft.

EVERSHED PORTABLE DIRECT READING.
OHMMETER WITH MAGNETIC GENERATOR.

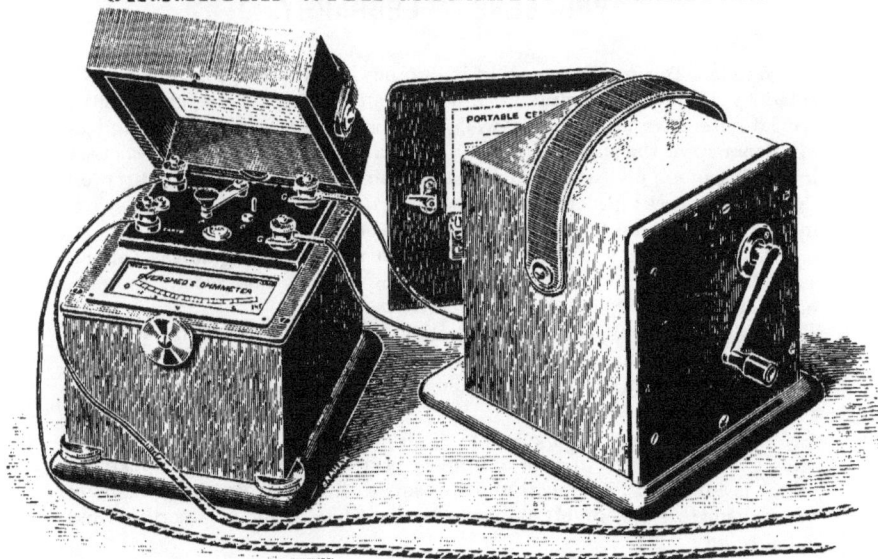

Fig. 106.

W6080. **Evershed Ohmmeter with Generator** $200 00
Range 0 to 50 megohms.
W6081. **Evershed Ohmmeter with Generator**......... 160 00
Range 0 to 10 megohms.
W6082. **Evershed Ohmmeter with Generator**. . 148 00
Range 0 to 5 megohms.
W6083. **Evershed Ohmmeter with Generator**. 160 00
Range 0 to 1000 ohms.
W6084. **Evershed Ohmmeter with Generator**......... 160 00
Range 0 to 100 ohms.

The Evershed Ohmmeter is direct reading, with an open scale and clear figures, and the Generator is of special construction, enabling high voltage to be obtained by turning the handle at a moderate speed. The reading of the Ohmmeter is quite independent of the speed at which the Generator is driven. All that is necessary to measure the resistance of any circuit is to couple it to the terminals of the Ohmmeter and turn the Generator handle, when the needle at once points to the value of the resistance of the scale.

This apparatus is manufactured in London, and will be imported to special order at above prices, which are f. o. b. New York, boxing extra.

MAGNETO BELLS FOR TESTING CIRCUITS.

Fig. 107.

W6088. **Portable Magneto** (with inside bells) .. .$9 00
   Rings through 10000 ohms.

W6089. **Portable Magneto** (with inside bells). .. 10 50
   Rings through 35000 ohms.

W6090. **Portable Magneto** (with inside bells). 12 50
   Rings through 50000 ohms.

SPEED INDICATORS.

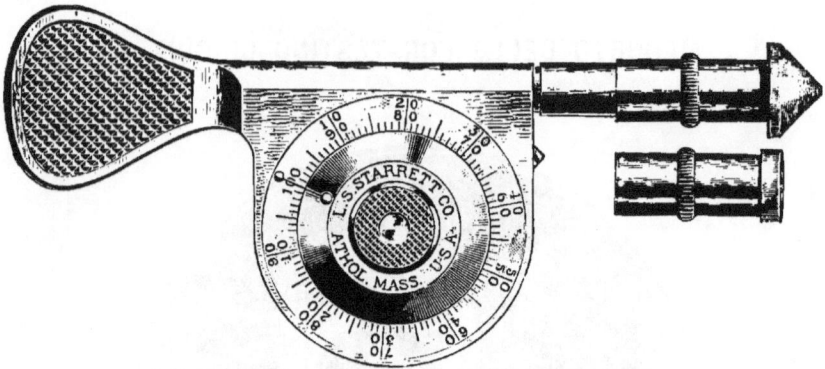

W6092. **Starrett's High-Speed Indicator No. 104** $1 00

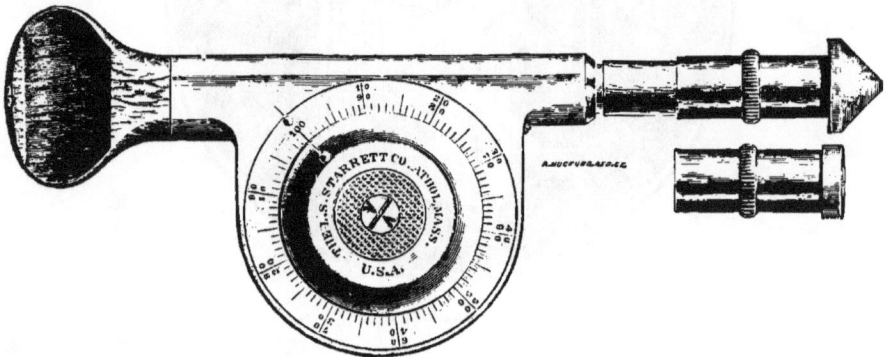

W6094. **Starrett's Improved Speed Indicator No. 100** ..$1 50

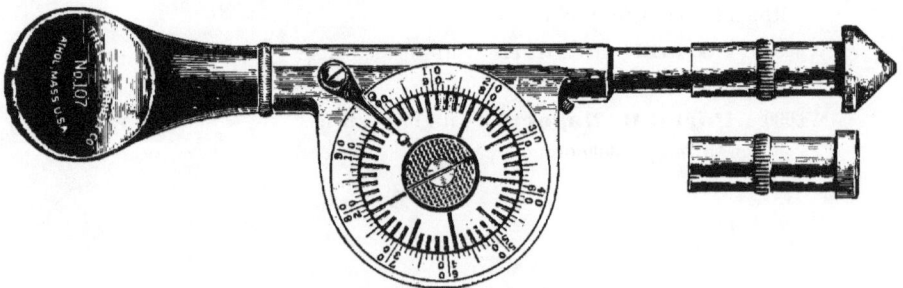

W6096. **Starrett's Registering Speed Indicator No. 107**.$3 00

TACHOMETERS

For Permanent Attachment.

Fig. 111.

W7000.   5½ inch dial, up to 2000 revolutions per minute. $55 00
W7001.   7½ inch dial, up to 2000 revolutions per minute.......... 60 00
W7002.   5½ inch dial, for more than 2000 revolutions per minute... 75 00
W7003.   7½ inch dial, for more than 2000 revolutions per minute... 80 00

These instruments have been designed for the purpose of ascertaining at a glance the number of revolutions made by rotating shafts. Their construction is based upon centrifugal power, and they consist of a case in which are mounted a pendulum ring, in connection with a fixed shaft, a sliding rod and an indicating movement.

The apparatus is very sensitive and will indicate the slightest deviation in speed. *All internal parts require no oiling.*

These Tachometers have been applied, with great success, to electric light engines, flour and cotton mills, and can be used to advantage on all machinery of which it is essential to know at all times the exact speed at which it is moving.

It is very important that the belt by which this instrument is driven be very smooth and pliable; an endless rubber belt answers the purpose best; if a joint is made it should be made without lapping.

The diameter of the driving pulley is marked on the dial.

The motion may be in either direction, and the apparatus may be placed in a vertical or horizontal position, or at any desired angle between these positions.

Tachometers of this class, with dials:

From about   30 to   300 revolutions per minute

        "   40 "   400   "   "

        "   50 "   500   "   "

        "   100 " 1000   "   "

        "   200 " 2000   "   "

can be furnished promptly; instruments for other speeds are made to order.
In ordering it should be stated for what speed the Tachometer is wanted.

PORTABLE TACHOMETERS.

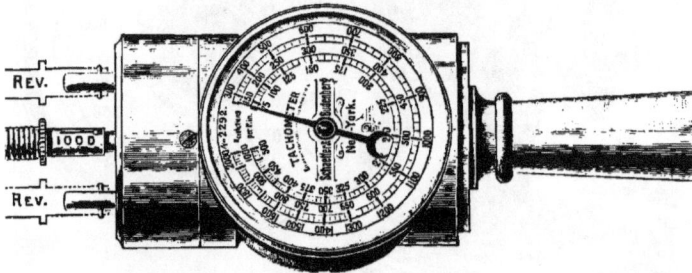

Fig. 112.

W7004.   3 inch dial, 50 to 2000 revolutions per minute. $55 00
W7005.   3 inch dial, 40 to 3000 revolutions per minute............. 60 00

    The Portable Tachometers are similar in construction to the permanent attachment form. By applying them to the center of rotating shafts they will instantly and correctly indicate the number of revolutions of the shaft per minute.

    They are of excellent workmanship and finish, are furnished in morocco leather cases for easy transportation, and are recommended for their adaptability for obtaining accurate indications of the variations in speed of slow and fast running machinery.

Diameter of Dial... .... 3 inches
Depth of Case.. 4   "
Entire Length 10   "
Total Weight, including morocco case...................... ....5 lbs.

    On unscrewing the end of handle a set-screw will be found, by means of which any lost motion may be taken up when necessary.

    W7004 has one scale from 100 to 1000 revolutions per minute.

    With this Tachometer are furnished 2 sets of geared wheels in proportion of 1 to 2, by means of which speeds between 50 and 500, respectively, 200 and 2000 revolutions may be indicated in such a way that when using the gear marked 500 the indications of the pointer must be divided by 2, and when using the gear marked 2000 the indications of the pointer must be multiplied by 2.

These gear wheels are easily attached to an arbor and held in position by a locknut provided for that purpose.

W7005 is provided with 3 scales, namely:

From 40 to 200 revolutions per minute
" 120 " 600 " "
" 600 " 3000 " "

The point only is detachable, and may be fastened on any of the 3 arbors, marked respectively 200, 600 and 3000 revolutions, and which correspond to the 3 scales on dial.

WESTON INSTRUMENTS.

No attempt has been made below to give a complete list of Weston instruments; their number is so great that a separate catalogue would be required. But it is thought that the list printed below will cover the average needs of the average user. To those not finding below exactly what they wish a more complete catalogue will be sent on application.

WESTON PORTABLE VOLTMETERS FOR DIRECT CURRENT CIRCUITS.

Fig. 120. WESTON PORTABLE AMMETER OR VOLTMETER.

CODE WORD.			PRICE.
W20. *Reprint.*	0 to 150 volts. 1 volt divisions, readable to $\frac{1}{16}$ v.		$55 00
W21. *Reprisal.*	0 to 150 " 1 " " " " $\frac{1}{10}$ "		
	Contact Key and Calibrating Coil.........		60 00
W22. *Reproach.*	0 to 150 volts. 1 volt divisions, readable to $\frac{1}{10}$ v.		
	0 to 5 " $\frac{1}{30}$ " " " " $\frac{1}{300}$ "		
	Contact Key............................		75 00

W23. *Reprove.* 0 to 150 volts. 1 volt divisions, readable to $\frac{1}{10}$ v.
0 to 3 " $\frac{1}{50}$ " " " " $\frac{1}{500}$ "
Contact Key........... .... 75 00

W24. *Reprune.* 0 to 150 volts. 1 volt divisions, readable to $\frac{1}{10}$ v.
0 to 15 " $\frac{1}{10}$ " " " " $\frac{1}{100}$ "
Contact Key............................ 75 00

W25. *Reptatus.* 0 to 300 volts. 2 volt divisions, readable to $\frac{1}{5}$ v.
0 to 150 " 1 " " " " $\frac{1}{10}$ "
Contact Key............................ 77 50

W26. *Reptile.* 0 to 300 volts. 2 volt divisions, readable to $\frac{1}{5}$ v.
Contact Key and Calibrating Coil......... 65 00

W27. *Republic.* 0 to 450 volts. 3 volt divisions, readable to $\frac{1}{3}$ v.
Contact Key and Calibrating Coil..... ... 70 00

W28. *Repulse.* 0 to 600 volts. 5 volt divisions, readable to $\frac{1}{2}$ v. 65 00

W29. *Requital.* 0 to 600 " 5 " " " " $\frac{1}{2}$ " 70 00
Contact Key and Calibrating Coil.........

W30. *Resemble.* 0 to 600 volts. 4 volt divisions, readable to $\frac{1}{2}$ v.
0 to 150 " 1 " " " " $\frac{1}{10}$ "
Contact Key 80 00

W31. *Reservoir.* 0 to 750 volts. 5 volt divisions, readable to $\frac{1}{2}$ v.
0 to 150 " 1 " " " " $\frac{1}{10}$ "
Contact Key 80 00

W32. *Reside.* 0 to 600 volts. 4 volt divisions, readable to $\frac{1}{2}$ v.
0 to 300 " 2 " " " " $\frac{1}{5}$ "
Contact Key 80 00

W33. *Residue.* 0 to 750 volts. 5 volt divisions, readable to $\frac{1}{2}$ v. 70 00

W34. *Resin.* 0 to 1500 " 10 " " " " 1 " 80 00
Reversing Key for any of above voltmeters,
extra. 2 50

PORTABLE MILLI-VOLTMETERS FOR DIRECT CURRENT CIRCUITS.

W34A. *Restful.* 0 to 0.02 volt. 100 divisions; each representing 0.0002 v.... .... $50 00

W34B. *Restrain.* 0 to 0.01 volt. 100 divisions, right and left;
each representing 0.0001 v.............. 50 00

W34C. *Resume.* 0 to 0.01 volt. 100 divisions, right and left,
contact key increasing sensibility ten
times 55 00

W34D. *Result.* 0 to 0.02 volt. 100 divisions; each representing 0.0002 v...................... ...... 55 00
0 to 0.2 volt. 100 divisions; each representing 0.002 v............................ 55 00

PORTABLE MILLI-VOLTMETERS WITH DETACHABLE SHUNTS.

These instruments indicate directly in *amperes*, and serve the same purpose as ammeters with one, two or more scales. The price of Milli-Voltmeter must be added to the price of a Shunt, as the instruments are used together.

W35. (*Resultant.*) Weston Portable Milli-Voltmeter.............. $50 00

SINGLE SHUNTS.

W36.	*Rab.*	0 to	1 a	$10 00
W37.	*Rabbit.*	0 to	3 a	12 50
W38.	*Rabble.*	0 to	5 a	15 00
W39.	*Rabies.*	0 to	10 a	15 00
W40.	*Rabinet.*	0 to	15 a	15 00
W41.	*Race.*	0 to	25 a	17 50
W42.	*Racme.*	0 to	30 a	20 00
W43.	*Racemic.*	0 to	50 a	22 50
W44.	*Racer.*	0 to	75 a	25 00
W45.	*Raceway.*	0 to	100 a	30 00
W46.	*Rachiel.*	0 to	150 a	32 50
W47.	*Rachis.*	0 to	200 a	35 00
W48.	*Rachitic.*	0 to	250 a	37 50
W49.	*Rack.*	0 to	300 a	40 00
W50.	*Racker.*	0 to	400 a	45 00
W51.	*Racket.*	0 to	500 a	50 00
W52.	*Rackety.*	0 to	600 a	55 00
W53.	*Racking.*	0 to	750 a	60 00
W54.	*Rackle.*	0 to	800 a	65 00
W55.	*Rackless.*	0 to	1000 a	80 00
W56.	*Racoon.*	0 to	1500 a	130 00

SHUNTS OF TWO RANGES COMBINED IN ONE CASE.

W57.	*Rad.*	0 to 1 0 to 5	a	$20 00
W58.	*Raddle.*	0 to 1 0 to 10	a	22 50
W59.	*Raddock.*	0 to 1.5 0 to 15	a	25 00
W60.	*Rade.*	0 to 3 0 to 15	a	25 00
W61.	*Radially.*	0 to 3 0 to 30	a	25 00
W62.	*Radian.*	0 to 4 0 to 40	a	27 50

W63.	*Radiance.*	$\begin{cases} 0\ \text{to}\ 5 \\ 0\ \text{to}\ 50 \end{cases}$ a..	30 00
W64.	*Radiant.*	$\begin{cases} 0\ \text{to}\ 15 \\ 0\ \text{to}\ 75 \end{cases}$ a..........................	35 00
W65.	*Radiate.*	$\begin{cases} 0\ \text{to}\ 10 \\ 0\ \text{to}\ 100 \end{cases}$ a......	40 00
W66.	*Radiation.*	$\begin{cases} 0\ \text{to}\ 15 \\ 0\ \text{to}\ 150 \end{cases}$ a...........................	42 50
W67.	*Radiator.*	$\begin{cases} 0\ \text{to}\ 30 \\ 0\ \text{to}\ 150 \end{cases}$ a...	45 00
W68.	*Radical.*	$\begin{cases} 0\ \text{to}\ 20 \\ 0\ \text{to}\ 200 \end{cases}$ a.......	47 50
W69.	*Radicate.*	$\begin{cases} 0\ \text{to}\ 15 \\ 0\ \text{to}\ 300 \end{cases}$ a.......................... ..	52 50
W70.	*Radication.*	$\begin{cases} 0\ \text{to}\ 30 \\ 0\ \text{to}\ 300 \end{cases}$ a.............................	55 00
W71.	*Radicose.*	$\begin{cases} 0\ \text{to}\ 150 \\ 0\ \text{to}\ 300 \end{cases}$ a	60 00
W72.	*Radio.*	$\begin{cases} 0\ \text{to}\ 40 \\ 0\ \text{to}\ 400 \end{cases}$ a.............................	60 00
W73.	*Radolite.*	$\begin{cases} 0\ \text{to}\ 200 \\ 0\ \text{to}\ 400 \end{cases}$ a.............................	65 00
W74.	*Radiolus.*	$\begin{cases} 0\ \text{to}\ 50 \\ 0\ \text{to}\ 500 \end{cases}$ a............................	62 50
W75.	*Radish.*	$\begin{cases} 0\ \text{to}\ 100 \\ 0\ \text{to}\ 500 \end{cases}$ a............................	65 00
W76.	*Radius.*	$\begin{cases} 0\ \text{to}\ 200 \\ 0\ \text{to}\ 500 \end{cases}$ a	65 00
W77.	*Radix.*	$\begin{cases} 0\ \text{to}\ 60 \\ 0\ \text{to}\ 600 \end{cases}$ a............................	65 00
W78.	*Radula.*	$\begin{cases} 0\ \text{to}\ 300 \\ 0\ \text{to}\ 600 \end{cases}$ a...............	80 00
W79.	*Radulas.*	$\begin{cases} 0\ \text{to}\ 300 \\ 0\ \text{to}\ 750 \end{cases}$ a........................	90 00
W80.	*Radulate.*	$\begin{cases} 0\ \text{to}\ 500 \\ 0\ \text{to}\ 1000 \end{cases}$ a....................	140 00

PORTABLE AMMETERS FOR DIRECT CURRENT CIRCUITS.

W81.	*Revere.*	0 to 5 amp.,	$\frac{1}{20}$ a. divisions,	readable to	$\frac{1}{200}$ a.	$65 00
W82.	*Revered.*	0 to 15 "	$\frac{1}{10}$ "	"	"	"	$\frac{1}{100}$ "..	65 00
W83.	*Reversal.*	0 to 25 "	$\frac{1}{4}$ "	"	"	"	$\frac{1}{40}$ "..	65 00
W84.	*Review.*	0 to 50 "	$\frac{1}{2}$ "	"	"	"	$\frac{1}{20}$ "..	65 00
W85.	*Revile.*	0 to 100 "	1 "	"	"	"	$\frac{1}{10}$ "..	70 00
W86.	*Revival.*	0 to 150 "	1 "	"	"	"	$\frac{1}{10}$ "..	75 00
W87.	*Revivalist.*	0 to 200 "	2 "	"	"	"	$\frac{1}{5}$ "..	80 00
W88.	*Revoice.*	0 to 250 "	2 "	"	"	"	$\frac{1}{5}$ "..	80 00
W89.	*Revolute.*	0 to 300 "	2 "	"	"	"	$\frac{1}{5}$ ".	80 00
W90.	*Rewake.*	0 to 400 "	4 "	"	"	"	$\frac{2}{5}$ "..	90 00
W91.	*Rewardless*	0 to 500 "	5 "	"	"	"	$\frac{1}{2}$ "..	90 00

WESTON PORTABLE MIL-AMMETERS FOR DIRECT CURRENT CIRCUITS.

W100. *Recticule.*	0 to	150	mil-amp., readable to	$\frac{1}{10}$	mil-amp	....	$50 00
W101. *Retinue.*	0 to	300	"	"	"	$\frac{1}{3}$	"	....	50 00
W102. *Retouch.*	0 to	600	"	"	"	$\frac{1}{2}$	"	....	50 00
W103. *Reuben.*	0 to	1000	"	"	"	1	"	....	50 00
W104. *Reunite.*	0 to	1500	"	"	"	1	"	....	55 00

W105. *Revel.* 0 to 500 " " " $\frac{1}{2}$ " $\Big\}$ .. 60 00
               0 to 50 " " " $\frac{1}{20}$ "

W106. *Revelry.* 0 to 500 " " " $\frac{1}{2}$ " $\Big\}$ .. 60 00
               0 to 15 " " " $\frac{1}{100}$ "

W107.*Revenge.* 0 to 500 " " " $\frac{1}{10}$ " $\Big\}$ .. 75 00
               0 to 10 " " " $\frac{1}{100}$ "

HIGH RANGE D. C. VOLTMETERS.

The Voltmeter is provided with a separate box, specially designed for high insulation, and containing extra resistance.

W110. *Resolute.* 0 to 2250 volts; with extra resistance $\Big\}$ $100 00
               0 to 150 " without " "

W111. *Resolve.* 0 to 3000 " with extra resistance $\Big\}$ 105 00
               0 to 150 " without " "

W112. *Resolvent.* 0 to 3750 " with extra resistance $\Big\}$ 115 00
               0 to 150 " without " "

W113. *Resonant.* 0 to 4500 " with extra resistance $\Big\}$ 125 00
               0 to 150 " without " "

W114. *Resound.* 0 to 5250 " with extra resistance $\Big\}$ 135 00
               0 to 150 " without " "

W115. *Restful.* 0 to 6000 " with extra resistance $\Big\}$ 145 00
               0 to 150 " without " "

---

* With resistance box of two values, which, when used in conjunction with the lower scale, indicates to 10 volts or 100 volts.

INSPECTOR'S STYLE D. C. VOLTMETERS.

Fig. 121. INSPECTOR'S D. C. VOLTMETER.

Under this designation any of the preceding voltmeters are furnished, permanently mounted, in a polished mahogany carrying box   (See fig. 121.) There is a reversing key, and contacts are made with socket-tipped flexibles attaching to studs, instead of by means of the usual binding posts. In connecting to lamp circuits the lamp is replaced by the proper "adapter" into which the flexible is arranged to fit.

W117. *Restinct.*	Mahogany Box, with lock and key and flexible.............. .. ....Add to list price,	$4 00
W118. *Restiness.*	Mahogany Box, with compartment for adapters and flexible.........Add to list price,	5 00
W119. *Restitute.*	Mahogany Box, with compartment for adapters and flexible, and four adapters for Edison, T. H., Westinghouse and Weston respectively............Add to list price,	10 00

WESTON PORTABLE VOLTMETERS FOR ALTERNATING AND DIRECT CURRENT CIRCUITS.

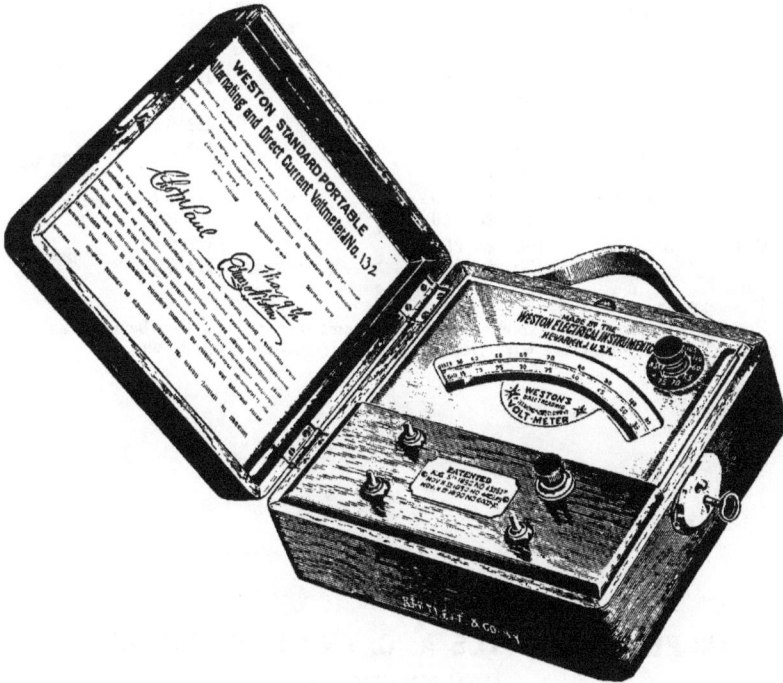

Fig. 122.

Each instrument is enclosed in a handsome leather case, supplied with shoulder strap, lock and key, and furnished with a pair of flexible connecting cords; for both of which no extra charge is made.

SINGLE SCALE INSTRUMENTS.

W120.	*Misgradito.*	1 5 to	7.5 volts.	$\frac{1}{20}$ v. divisions	....	$60 00
W121.	*Misperava.*	1.5 to	10	"	$\frac{1}{10}$ "	"	....	60 00
W122.	*Misnregio.*	2 to	12	"	$\frac{1}{10}$ "	"	....	60 00
W123.	*Misturaste.*	3 to	15	"	$\frac{1}{10}$ "	"	....	60 00
W124.	*Misturarel.*	3 to	20	"	$\frac{1}{10}$ "	"	....	60 00
W125.	*Misuriamo.*	10 to	60	"	$\frac{1}{2}$ "	"	....	60 00
W126.	*Misvengono.*	15 to	75	"	$\frac{1}{2}$ "	"	....	60 00
W127.	*Misven.*	20 to	120	"	1 "	"	....	65 00
W128.	*Miterone.*	30 to	150	"	1 "	"	....	67 50
W129.	*Mitescant.*	60 to	300	"	2 "	"	....	70 00
W130.	*Mitichero.*	100 to	600	"	5 "	"	....	80 00

DOUBLE SCALE INSTRUMENTS.

W140.	*Mobiliter.*	3	to	15	volts, divided like	B213	} · · ·	$70 00
		1.5	to	7.5	" " "	B210		
W141.	*Mocadero.*	3	to	20	" " "	B214	} · · ·	70 00
		1.5	to	10	" " "	B211		
W142.	*Moccolino.*	20	to	120	" " "	B217	} · · ·	70 00
		10	to	60	" " "	B215		
W143.	*Moccabaic.*	30	to	150	" " "	B218	} · · ·	75 00
		15	to	75	" " "	B216		
W144.	*Mocctonas.*	30	to	200	" " "	B214	} · · ·	75 00
		15	to	100	" " "	B211		
W145.	*Mochadura.*	60	to	300	" " "	B219	} · · ·	80 00
		30	to	150	" " "	B218		
W146.	*Mochiler.*	100	to	600	" " "	B220	} · · ·	90 00
		50	to	300	" " "	B219		

* The higher range is compensated for temperature, and can be used as an Ammeter, in combination with Detachable Shunts (pages 149 and 150).

INSPECTOR'S STYLE A. C. AND D. C. PORTABLE VOLTMETERS.

W147. *Adhere.* Voltmeters W120 to W146, with compartment for flexible cords.....Additional to list price, $2 50

W148. *Adhered.* Voltmeters W120 to W146, with lamp adaptors for Edison, Thomson-Houston, Westinghouse and Weston sockets. Also flexible cord with connecting plug......Additional to list price, 7 50

Fig. 123.

WESTON PORTABLE WATTMETERS FOR ALTERNATING AND DIRECT CURRENT CIRCUITS.

Each instrument is furnished with a carrying case of polished cherry. For No. W164 and over, this case contains a separate compartment for linen-covered flexible cables, which are supplied with the instrument.

	Range				Max Current.	Max Voltage.	Price
W160. *Adrift.*	0 to	150 watts.	1 watt div.		2 amp.,	150 v.	$70 00
W161. *Adroit.*	0 to	75 "	½ "	"	2 "	75 " }	75 00
	0 to	150 "	1 "	"	2 "	150 " }	
W162. *Adroitness.*	0 to	1500 "	10 "	"	10 "	150 "	70 00
W163. *Advance.*	0 to	750 "	5 "	"	10 "	75 " }	75 00
	0 to	1500 "	10 "	"	10 "	150 " }	
W164. *Advantage.*	0 to	3750 "	25 "	"	25 "	150 "	75 00
W165· *Adversary.*	0 to	1500 "	10 "	"	25 "	75 " }	80 00
	0 to	3000 "	20 "	"	25 "	150 " }	
W166. *Adverse.*	0 to	7500 "	50 "	"	50 "	150 "	75 00
W167. *Adversely.*	0 to	3750 "	25 "	"	50 "	75 " }	80 00
	0 to	7500 "	50 "	"	50 "	150 " }	
W168. *Advice.*	0 to	15000 "	100 "	"	100 "	150 "	80 00
W169. *Advisable.*	0 to	7500 "	50 "	"	100 "	75 " }	85 00
	0 to	15000 "	100 "	"	100 "	150 " }	
W170. *Advisably.*	0 to	30000 "	200 "	"	200 "	150 "	85 00
W171. *Advisedly.*	0 to	15000 "	100 "	"	200 "	75 " }	90 00
	0 to	30000 "	200 "	"	200 "	150 " }	
W172.*Adroitly.*	0 to	150 "	1 "	"	2 "	150 "	85 00

* For Inspectors, Sales Agents and Manufacturers of incandescent lamps. Furnished with adaptors for Edison, Westinghouse and Thomson-Houston sockets, and flexible connecting cords.

MULTIPLIERS FOR WESTON PORTABLE WATTMETERS FOR ALTERNATING AND DIRECT CURRENT CIRCUITS.

The capacity in *volts* only is increased; the maximum current capacity remains unchanged.

		Multiplier of	Maximum Voltage.	For W160–W163.	For W164–W165.	For W166–W171.
W180.	*Minutaglia.*	2	300 volts.	$12 00	$15 00	$15 00
W181.	*Miratrix.*	4	600 "	17 50	20 00	20 00
W182.	*Mirrastes.*	5	750 "	20 00	22 50	22 50
W183.	*Martidano.*	10	1500 "	30 00	35 00	35 00
W184.	*Misaltato.*	15	2250 "	35 00	40 00	60 00
W185.	*Miscadere.*	20	3000 "	40 00	60 00	70 00
W186.	*Miscidato.*	25	3750 "	60 00	70 00	80 00
W187.	*Miscredono.*	30	4500 "	65 00	75 00	85 00
W188.	*Misericors.*	40	6000 "	75 00	90 00	100 00

For multipliers subdivided to give intermediate ranges, add $5.00 for each step.

PORTABLE AMMETERS, VOLTMETERS AND WATTMETERS.

(Thomson Inclined Coil Type.)

These instruments are accurate and permanent for alternating currents of any frequency and give *fairly* good results with direct currents. They are strongly made and will stand the usage of the ordinary "inspector" and of shop practice admirably.

Wood cases are provided for all except the "pocket" instruments.

AMMETERS.

	Capacity in Amperes.	Price.
W200.	2	$40 00
W201.	10	40 00
W202.	15	40 00
W203.	25	40 00
W204.	50	40 00
W205.	100	50 00
W206.	200	50 00

VOLTMETERS.

	Capacity in Volts.	Price.
W207.	65	$50 00
W208.	130	50 00
W209.	300	50 00
W210.	600	50 00

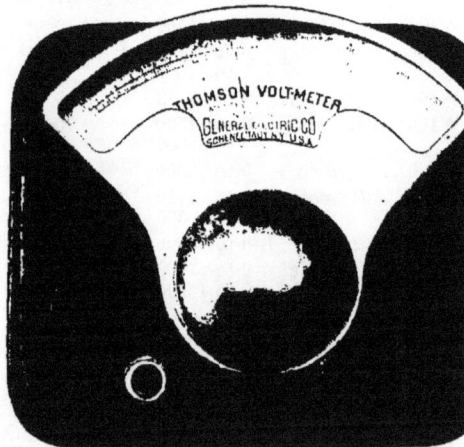

Fig. 124. INCLINED COIL VOLTMETER.

MULTIPLIERS
for Inclined Coil Portable Voltmeters.

THESE DOUBLE THE RANGE GIVEN.

	Capacity of Voltmeters in Volts.	List Price.
W211.	65	$20 00
W212.	130	20 00
W213.	600	25 00

WATTMETERS.

	Capacity in Watts.	Price.
W213A.	150	$75 00
W213B.	300	75 00
W213C.	1500	75 00
W213D.	2500	75 00

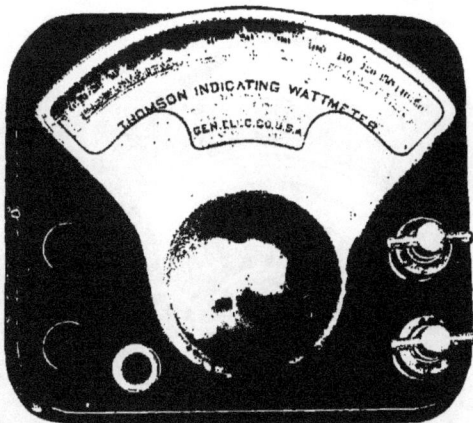

Fig. 125.

Inclined Coil Indicating Wattmeters will be found particularly useful for incandescent lamp, transformer and alternating arc lamp measurements. They are accurate even upon circuits carrying an inductive load such as motors or economy coils. All sizes may be used at any voltage up to 150 volts, and the permissible maximum current is marked upon each instrument. Indicating Wattmeters with special resistances for higher voltages can be supplied on special order. The 150 watt instrument is useful for measuring iron losses of transformers and for incandescent lamp work. The Indicating Wattmeter gives accurate results for direct current measurements, provided two observations are taken for each reading. The connections of the instrument, both current and voltage, must in this case be reversed between the observations, and the mean of the two observations taken as the current indication.

A special wattmeter can be made on order for balanced three-phase circuits.

A portable wattmeter with special facilities for lamp testing has recently been designed. It is known as the Lamp Inspectors' Indicating Wattmeter and is described below.

LAMP INSPECTORS' WATTMETERS.

Fig. 126. LAMP INSPECTORS' WATTMETER.

Particularly recommended for measuring watts used in lamps or small motors; may be used on either direct or alternating circuits. Has a flexible with adapters to fit Edison, Westinghouse or T. H. system. Capacity, 150 watts. Mounted in polished case with snap lock. Measures over all, 7⅜ x 7⅜ x 4½ thick.

W214. **Lamp Inspectors' Wattmeter**....................$100 00

POCKET AMMETERS AND VOLTMETERS.

Fig. 127.

These are strongly made instruments, small enough to be carried in the coat pocket. The scale is about 3½" long.

AMMETERS.

	Capacity in Amperes.	List Price.
W215.	2	$25 00
W216.	10	25 00
W217.	25	25 00

VOLTMETERS.

	Capacity in Volts.	List Price.
W218.	75	$30 00
W219.	150	30 00

W225. "Standard" Electric Gauge $6 00

For making comparative tests of battery E. M. F. and current. Size of an ordinary watch and finished in nickel. Resistance about 10 ohms.

W226. "Standard" Pocket Voltmeter $7 00

Same as W225 but graduated directly in volts with a maximum of 2 volts. A very convenient instrument for those having charge of storage batteries.

Fig. 128. W225.

"STUDENTS" ELECTRICAL APPARATUS

FOR

COLLEGE AND HIGH SCHOOL LABORATORIES.

Unless otherwise stated the general specifications for apparatus described on pages 160 to 165 are:—
Wood-work—Cherry, with shellac finish.
Metal-work—Lacquered brass or copper with all castings rough and blacked.
All key contacts are of platinum.

KEYS.

F500.	**Single Contact Key**	$1 75
F501.	**Double Contact Key**	2 50
F502.	**Single Contact and Short Circuit Key**	2 50
F503.	**Discharge Key**	4 00
F504.	**Reversing Key**, improved form	4 50
F505.	**Pohl Commutator**	2 25

CONDENSERS.

Of mica and accurate within, approximately, 2%.   The brass terminal blocks mounted upon hard rubber strips for insulation.

F511.	**1-3 Mica-farad**		$22 50
F512.	**1-2** "		25 00
F513.	**1** "		30 00
F514.	**1** "	**sub-divided**	40 00

In five sections of 0.05, 0.05, 0.2, 0.2, and 0.5 M. F's Segments are connected between parallel brass blocks like the coils in a "Multiple arc" resistance box; thus allowing of either series or multiple arrangement, or combination of series and multiple. (*Same construction exactly as fig. 7.*)

Fig. 135. F529.

GALVANOMETERS.

F520 Simple Detector Galvanometer.....................$4 50

After design of Prof. E. A. Partridge, Central Manual Training School, Philadelphia. Glass pointer over 4" cardboard scale, divided in degrees. System hangs upon a glass jewel and cannot get displaced for any position of pointer. Deflects with $\frac{1}{30000}$ ampere of current.

F521. Astatic Galvanometer $8 00

Glass pointer over 4" cardboard scale, divided in degrees: fibre suspension 6" long; glass bell jar protects from dust and air currents.

F522. Reflecting Astatic Galvanometer...................$9 00

Harvard design as modified by us. Pointer is of glass, moving over 4" cardboard degree scale; and hung by a silk fibre. Coil box fits upon a brass base casting in which is a central cup to be filled with oil or turpentine. The needle system ends below in an aluminum vane, which dropping into this cup, effectively damps the motion of the needle. The glass top carries the suspension tube and is readily removable thus making replacement of the suspension a simple matter when necessary.

F523. Kelvin Galvanometer...............................$25 00

Resistance about 5000 ohms (coils in series).

F524. Kelvin Galvanometer............................... 20 00

Resistance about 1000 ohms (coils in series).

F525. Kelvin Galvanometer............................... 18 00

Resistance about 100 ohms (coils in series).

Usual form with four coils. Each coil has its terminals joined to a pair of binding posts upon a hard rubber plate itself fixed to the base of the instrument, so that the coils may be joined up in series or multiple or in any desired combination of series and multiple. The fibre suspension is about 3 inches long and its support bears a "scissors" control magnet for altering the sensibility. The frame of the instrument is of well dried wood and plainly but neatly finished. The front is hinged, to open up the system for inspection or repair of suspension.

F528. **Tangent Galvanometer**..............................$12 50

Coil frame of well dried cherry—mean diameter about 7⅜ inches. Has three sets of windings of 1, 5, and 10 turns, respectively, arranged for use separately or in series. Has glass pointer over 3″ cardboard scale, divided in degrees. System pivoted on glass jewel and cannot become displaced for any position of the instrument.

F529. **D'Arsonval Pointer and Reflecting Galvan-**
ometer...$12 50

Resistance about 1500 ohms, sensitiveness, 200-300 megohms Has magnet of tungsten steel and clamp to hold coil in position and allow of instrument being transported without risk of breakage.

Fig. 136.

RESISTANCE COILS, WHEATSTONE BRIDGES, ETC.

F533. **Standard One Ohm Coil**................ ....$5 00

With both binding posts and copper strip terminals (to fit gap in meter bridge). Accurate within ⅛%.

F534. **Standard Ten Ohm Coil**...........................$6 00

Accuracy, mounting, etc., the same as F533.

F535. **Resistance Box**.....................................$20 00

111.5 ohms divided into 10 coils of 0.5, 1, 1, 2, 3, 4, 10, 20, 30, and 40 ohms. Box of cherry with hard rubber top. Accuracy within ½%.

F536. **Resistance Box**.............................. ....$27 50

Same as F535, but with addition of coils of 100, 200, 300, 400, and 1000 ohms; 2,111.5 ohm total resistance.

F537. **Resistance Box**......................................$32 50

Same as F536, but with further additions of coils of 2,000, 3,000 and 4,000 ohms; 11,111.5 ohms total resistance.

F538, **Resistance Box and Wheatstone Bridge**........$40 00

Consists of resistance coils as in F537 with addition of bridge-arms of 10, 100, 1000 ohms on each side. Battery and galvanometer keys are mounted upon rubber top as part of the set. Accuracy of bridge-arms ⅛% and of coils ½%.

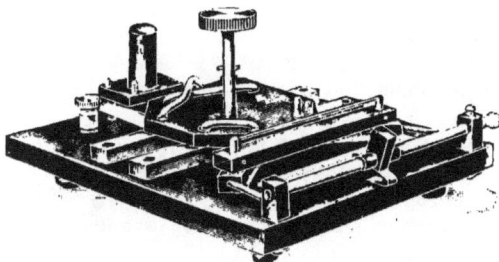

Fig. 137. F539.

F539. Carey-Foster Bridge and Commutator..........$35 00

Simplified form, complete with ratio coils of 1, 10, and 100 ohms. In design it resembles our high grade, model W4240, but is much more plainly and simply made, and has no battery commutator.

F545. Reading Telescope and Scale.....................$10.00

This is a thoroughly good and substantial instrument. The telescope-proper is exactly the same as is used in our " Aone " D'Arsonval Galvano-meter (see also W4220), and optically perfect. The stand is of cherry and solidly put together. This instrument has all necessary adjustments.

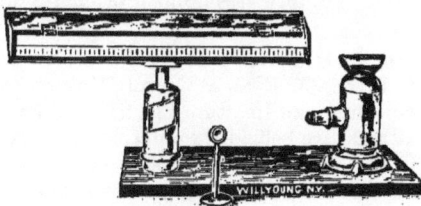

Fig. 138. F546.

F546. Lamp and Scale $17 50

Simple form for use with plane mirror. Either oil, gas, or electric as ordered. (Same as W4210, fig. 23, page 33.)

Fig. 139. F547.

F547. Magnetometer...................................... $16 00

Simple form suitable for students of the High School grade. Is strongly made and capable of giving results of considerable accuracy. Complete with deflecting and deflection magnets, inertia ring, etc.

F548. Copper Voltameter................................. $7 50

As designed and used by Prof. E. S. Ferry, of the University of Wisconsin. The anode plate is supported by a hard rubber plug, itself fitting into a hard rubber top from which it may be readily withdrawn without disturbing the cathode; a long slot in the top permits its passage. Cathode plate measures $3''x2\frac{3}{8}''$.

F550. Copper Voltameter for Heavy Currents........ $15 00

As described by Mr. A. W. Meikle and used in the University of Glasgow (See *Electrolysis of Copper Sulphate in Standardizing Electrical Instruments*—Proc. Physical Society of Glasgow University—January 27th, 1888 ; also see *Gray's Absolute Measurements in Electricity and Magnetism*, Vol. II., part II., page 419). Has glass cell about $8x9\frac{1}{2}''$. Five anode plates and four cathode plates, each $8\frac{1}{4}x10''$; each plate has two tongues at top which support it in a slot of the top framing ; connections being automatically made so soon as tongues drop into the slot. Plates are thus readily removable for weighing. With all the plates, currents up to 60 amperes are measurable, but any less number of plates may be used if desired.

F552. Ryan Spiral Coil Voltameter....... .... $7 50

As used by Prof. Harris T. Ryan at Cornell University. (See paper by Prof. Ryan before American Institute Electrical Engineers May 22d, 1889). Anode and cathode are helically wound wires, suspended vertically in the electrolyte, the cathode within the anode. The advantages claimed by Prof Ryan for this form of voltameter are : freedom from sharp edges and corners, and consequently greater firmness of deposit ; facility of cleaning by merely clamping one end of wire in vice, straightening, and sand-papering down. Electrodes are hung from a hard rubber plate, itself fixed to a telescopic supporting tube, so that by raising or lowering, the current may be kept perfectly constant against variations of E. M. F. The cathode also has an adjustment which allows it to be centered with reference to the anode.

F553. Earth Inductor...................................... $35 00

Coil frame about $10''$ in diameter, mounted in rectangular frame, the latter turning upon a horizontal axis and the former upon an axis at right angles to it. The whole is mounted upon a base with screws and bob for leveling. A special device throws the coil at uniform speed through 180° or allows of continuous rotation as desired. The wire is wound in layers, the diameter, etc., of which are exactly determined and furnished with each instrument.

F554. Electrometer............................... .. .. $17 50

Simple quadrant type, capable of good work and well insulated. All parts are accessible, and various sensibilities are easily obtained by a change of suspension ; gives deflections for a single cell of battery.

ELECTRO-DYNAMOMETERS.

These instruments are strongly but very plainly made, and may be depended upon for constancy and accuracy. The constant of the instrument is furnished in each case, but no calibration card or curve.

F560. Range 0.2 to 4 amperes................. .... $17 50

F562.   "   1 to 20   " 20 00

F564.   "   3 to 60   " 22 50

OSCILLOGRAPHS

AND THEIR USES.

The Oscillograph may be defined as an instrument whose deflection is proportional to the current flowing through it at each and every successive instant of time. The practical uses of such instrument are many since they record with great accuracy time changes of E. M. F. and current. Thus the simultaneous P. D. and current changes on making and breaking an inductive circuit, the charge and discharge curves of condensers, P. D. and current changes in a dynamo armature, etc., may all be directly recorded.

Such rapid variations as take place during the hissing of the arc lamp are faithfully depicted. Wave forms, phase difference, self induction, capacity, power factor, etc. are thus all rapidly obtained.

The Oscillograph puts a new and powerful weapon of research into the hands of electrical engineers.

HOTCHKISS OSCILLOGRAPH.

An apparatus for recording a variable current curve or two such curves simultaneously. This apparatus has been developed by Prof. Hotchkiss of Cornell University during the past few years. It was exhibited before the American Association for the Advancement of Science in 1896. (see *Elect. World*, Sept. 26, 1896.) The original form was described in the *Physical Review* of July, 1895. Other articles may be found in the same paper for March and September, 1896. Also in *Trans-*

actions A. I. E. E., vol. 13, May 20, 1896. The final and most elaborate article on the subject in which are given a number of full size curves will be found in the *Physical Review* for March, 1899. For an abstract of this article see the *Electrical World*.

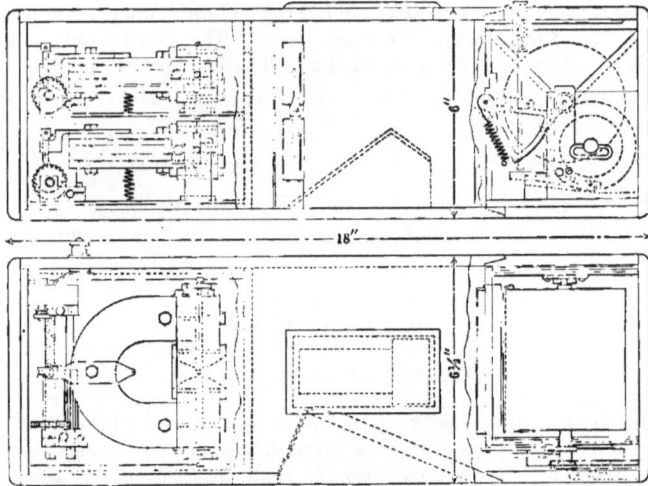

Fig. 140. HOTCHKISS OSCILLOGRAPH.

The instrument is portable and gives continuous, simultaneous records of two A. C. curves whether cyclic or non-cyclic so as to show their corresponding values for each instant of time.

The corresponding values of any number of cyclic curves may be obtained by taking two, then one of these with a third, and so on.

Essentially there are two Deprez Galvanometers which have a very minute moving system of soft iron, supported by a quartz fibre in the concentrated field of a permanent magnet. These are mounted one above the other in one end of a case measuring about 6 x 6½ x 18″ long (see fig. 140) In the other end of the case is a revolving drum carrying a photographic film (see figs. 141 and 142)

Sun light or arc light is seen through two narrow slits and is reflected from the mirror of the two needles upon a horizontal slit in the drum boxes.

The two currents to be recorded are joined to the deflecting coils of the two galvanometers. The mirror will then deflect on the closure of this current and since the film is passing before this slit the motion of the mirror will be recorded as a connecting photographic curve.

The needles are made so small that a frequency up to 8,000 to 10,000 double vibrations per second can be obtained. At the upper speed the needle may be forced to "overthrow" its legitimate swing, but, if so, the amount of this "overthrow" is very slight, so that the general curve may be taken as correct.

Fig. 141: HOTCHKISS OSCILLOGRAPH (Assembled).

Fig. 142. HOTCHKISS OSCILLOGRAPH (Not Assembled).

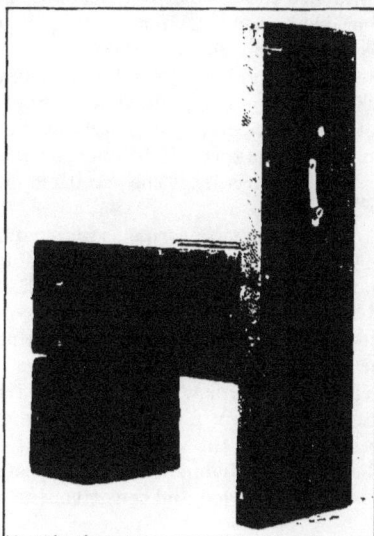

Fig. 143. SHOWING SLIDE WAY.

The galvanometer coils may be connected in series or in parallel and there are several sets of coils of different resistances so that one may be exchanged for another.

This instrument is also made with a "slide way" for dry plates (fig. 143). The plate is shot downward by a strong spring and then released by a special trigger having its energy taken up by an air cushion at the bottom.

We have been appointed by Prof. Hotchkiss sole licensees to manufacture this instrument, the designs being submitted to him for approval and correction and his advice will be constantly at our service during the construction of this instrument. The needles for each instrument will be made by him.

Further particulars and price will be furnished by us upon application.

THE DUDDELL OSCILLOGRAPH.

This instrument is very simple, consisting essentially of a strong magnetic field with pole pieces so shaped as to concentrate the lines of force into a thin vertical sheet. In this field are stretched two metal strips, which are held in tension, current passing up one of these strips and down another. A mirror attached to both strips deflects as the one strip moves forward and the other backward. Owing to the tension the period of system is exceedingly short, viz., about 1/10000 of a second. It is at the same time very "dead beat" and has a very low resistance and practically no self-induction. The sensitiveness, scale at 50 centimeters from mirror, is nearly 290 millimeters per ampere. The magnetic field is very intense and produced by an electro-magnet; it is energized by direct currents from 25, 50, 100 or 200 volt circuits. The exciting current is about ¼ ampere at 110 volts.

The instrument is made in two forms: single and double. In the latter there are two systems in one magnetic field (see figure 144). Hence, simultaneous observations may be made of any two E. M. F.s or currents, or of a current and an E. M. F. A fixed mirror gives the datum line. The movement of the spot of light can either be photographed on a moving plate or film or observed in a rotating mirrror. When the variations are periodic a vibrating mirror with the axis at right angles to that of the Oscillograph and operated by a synochronous motor will reflect the spot of light to a screen on which the stationary bright curve can be easily traced or photographed. The instrument has an adjustment for zero and and for slightly increasing the period and sensitiveness.

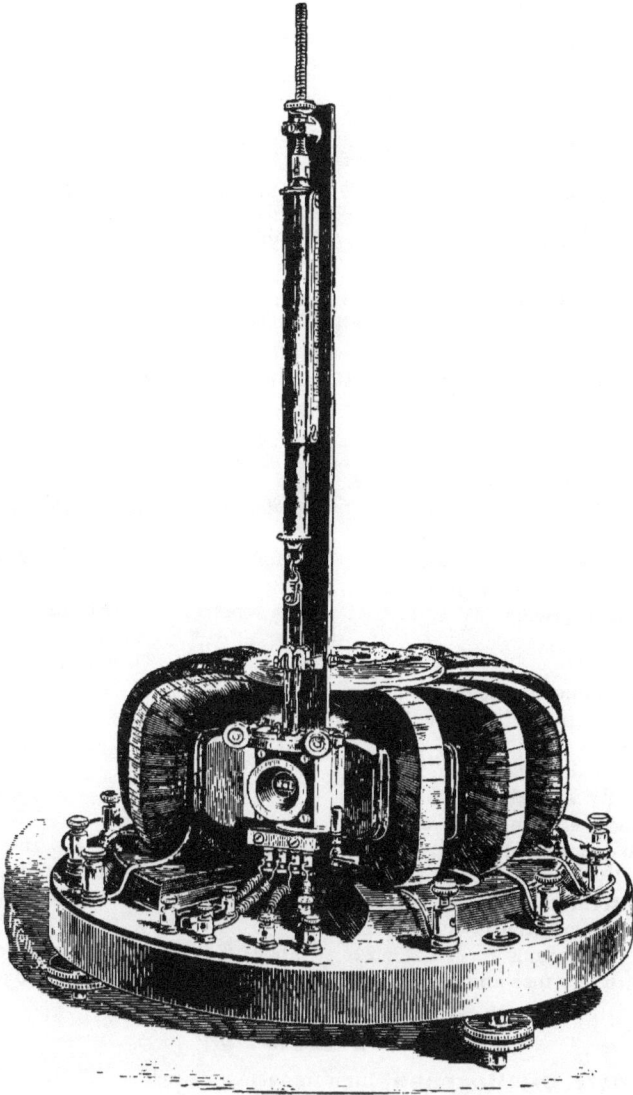

Fig. 144. DUDDELL OSCILLOGRAPH.

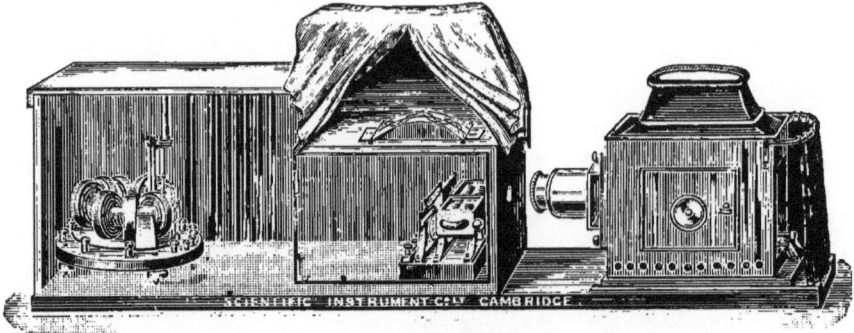

Fig. 145. DUDDELL OSCILLOGRAPH, WITH LANTERN AND MIRROR FOR
STATIC CURVES.

APPROXIMATE DATA OF THE DUDDELL OSCILLOGRAPHS.

	High Frequency Pattern.	Projection Pattern.
Resistance of Field Coils in series at 15° C	360 ohms	180 ohms
Normal Exciting Current with the coils in series	0.25 amp.	0.5 amp.
Normal tension on strips in Single Oscillograph	4 ozs.	This instrument is not made in the projection pattern.
Do. in Double Oscillograph	8 ozs.	6 lbs.
Periodic Time (undamped) with the above tension	$\frac{1}{8000}$ to $\frac{1}{10000}$ sec.	$\frac{1}{1800}$ to $\frac{1}{9000}$ sec.
Normal scale distance	50 cm.	300 cm.
Sensibility with the above tension, normal exciting current and scale distance	290 mm. per ampere	0.5 ampere
Normal Working Current in strips for alternate current wave forms	0.05 to 0.10 ampere	0.5 ampere
Resistance of strips without fuse and connections	about 2 ohms	about 1 ohm
Do. with one fuse and connections	about 10 ohms	about 3 ohms

PRICES.

1. **Single Oscillograph,** high frequency pattern............$187 50
With constants described on page 9.

2. **Double Oscillograph,** do...............................$225 00

3. **Single Oscillograph,** high frequency pattern............$300 00
Complete with tracing desk, synchronous motor, etc.

4. **Double Oscillograph,** high frequency pattern.... ......$337 50

Complete with tracing desk, synchronous motor, etc.

5. **Double Oscillograph,** high frequency pattern..........$487 50

Complete with revolving cylinder for photographically recording the wave forms with automatic exposing shutter and contact.

Interested parties are requested to correspond with us regarding this instrument and we shall be pleased to send such a special pamphlet giving fuller details. Colleges and scientific institutions may import this instrument free of duty and quotations will be made on request.

BOOKS AND PERIODICALS.

Being often asked by customers to recommend or secure for them books relating to such branches of electrical work as we are supposed to be experts in, we have prepared the following list. This list is not a large one and very possibly we have omitted some very good and important works. But we feel that all those mentioned are of the very best and to be heartily recommended.

WORKS ON THE X-RAY.

THE ABC OF THE X-RAY; by *W. H. Meadowcroft*—a brief general treatment of the whole subject..................... ...paper, $0 50
cloth, 75

THE X-RAY; by *Dr. W. T. Morton and Mr. W. J. Hammer*.. paper, 50
cloth, 75

ROENTGEN RAYS AND PHENOMENA OF THE ANODE AND CATHODE; by *Thompson & Anthony*—a consecutive historical treatment of the general phenomena of discharges in vacuum tubes..... 1 50

RADIOGRAPHY AND X-RAYS; by *S. R. Bottone*—contains details as to construction of apparatus, etc......... .... 1 00

RADIATION; by *Hyndmann*—a complete treatment of the subject of radiation in the ether 1 60

PRACTICAL RADIOGRAPHY; by *Menthal & Ward*................... 1 00

ROENTGEN RAYS; being the memoirs, of *Roentgen, Stokes* and *J. J. Thomson*—edited and arranged by *Prof. G. F. Barker*........ 60

ELECTRICAL TESTING AND MEASUREMENT.

ELECTRICAL MEASUREMENTS; by *Carhart & Patterson*—covers the
subject of laboratory measurement very nicely $2 00
ABSOLUTE MEASUREMENTS IN ELECTRICITY AND MAGNETISM; by
A. *Gray*—in two volumes . . . . { Vol. I, 510 pages,   3 25
                               { " II, 868 "   6 25
                               { Abridged edition,   1 25
HANDBOOK OF ELECTRICAL TESTING; by *H. R. Kempe*—a very
exhaustive work on laboratory methods and instruments, par-
ticularly as applied to submarine cable work; 5th edition. . . .   7 25
PHYSICAL MEASUREMENTS; by *Kohlrausch*—contains appendix on
electrical units and measurements 4 00
THE GALVANOMETER; by *Nichols*—being lectures to Cornell students.
Treats almost entirely of the Kelvin instrument 1 00
MEASUREMENT OF ELECTRICAL RESISTANCE; by *Price* 3 50
ELECTRICITY AND MAGNETISM (being vol. II of "Elementary Prac-
tical Physics"); by *Stewart & Gee*—is very good on the sub-
ject of current, E. M. F. and resistance measurements 2 25
PRACTICAL ELECTRICITY AND MAGNETISM; by *Henderson* 2 00
PRACTICAL ELECTRICITY; by *W. E. Ayrton*—covers whole field
briefly 2 50
PHOTOMETRY; by *Palaz* and translated by *Prof. Geo. W. Patterson* . .

PERIODICALS.

Per Annum.

ELECTRICAL WORLD AND ENGINEER, New York (weekly) $3 00
AMERICAN ELECTRICIAN, New York (monthly) 1 00
ELECTRICIAN, London (weekly) 8 00
ELECTRICAL ENGINEER, London, (weekly) 4 50
ELECTRICAL REVIEW AND TELEGRAPHIC JOURNAL, London (weekly)   5 75
L'ECLAIRAGE ELECTRIQUE, Paris, (weekly) 15 50
ELEKTROTECHNISCHE ZEITSCHRIFT, Berlin (weekly) 6 80

Special thanks are due the Electrical Engineer Institute of Correspondence Instruction for diagram cuts loaned.

ELMER G. WILLYOUNG.